Culture of Control

FARMING WITH THE SHACKLES OF THE CANADIAN WHEAT BOARD

Ken Motiuk

BY KEN MOTIUK

◆ FriesenPress

One Printers Way
Altona, MB R0G 0B0
Canada

www.friesenpress.com

Copyright © 2025 by Ken Motiuk
First Edition — 2025

This book is a recollection of the family legacy of the Motiuk family and, specifically, Ken Motiuk's journey. It pays tribute to all farmers who pioneered the land in Western Canada and built their homes and businesses there—not because of, but despite, the Canadian Wheat Board. Any errors or omissions in the text are the responsibility of the author.

ISBN
978-1-03-831754-4 (Hardcover)
978-1-03-831753-7 (Paperback)
978-1-03-831755-1 (eBook)

1. HISTORY, CANADA, PROVINCIAL, TERRITORIAL & LOCAL, PRAIRIE PROVINCES (AB, MB, SK)

Distributed to the trade by The Ingram Book Company

DEDICATION

To Wendy, my partner and best friend, who has been supportive and encouraging throughout. She kept the home fires burning the many years I was away promoting constructive change. She looked after our family and kept the farm going while maintaining her own profession as a nurse. Wendy's comments, suggestions, review, and edit of this narrative made it much easier to complete.

To our daughters and their families. May they forever appreciate, embrace, and maintain freedom and liberty in their lives and livelihoods. For the betterment of all, individuals must be able to exercise their independent knowledge, thoughts, and abilities in an environment unencumbered by trivial, arbitrary, and unreasonable bureaucratic zealousness.

TABLE OF CONTENTS

FOREWORD

Let me begin by thanking Ken for this insightful accounting of the many trials incurred by Western Canadian farmers foisted on them by the Canadian Wheat Board.

These pages are an accurate decades-long analysis of a government policy that has outlived its usefulness. The growing concern of these same farm families as to the pall cast over their economic well-being has been well documented.

The farm gate has historically been forced to become price takers. Sales of products were made at times and values not in consultation with farmers. Non-board commodities grew in volume and value but were still constrained by questionable logistics of the board. Commodities moved at the whims of the board, which controlled railcar allocation. Sales were lost, pricing was negatively affected, and buyers moved on to more stable suppliers.

Farmers embraced new commodities and moved en masse to canola and pulses—non-board crops that did not require the sacrosanct permit book and single point of delivery.

Farmers braved arrest and prosecution for going offside with these antiquated, government-enforced policies. The die was cast and from then on, the movement away from the past towards a brighter future began in earnest.

Thank you to those few who did so much.

Thank you to Ken for capturing it so eloquently.

Freedom is never free.

The Honourable Gerry Ritz, who served as Canada's minister of agriculture from 2007 thought 2015 under Prime Minister Stephen Harper.

≈

"Finally! Someone has stepped forward to tell the inside story of the management and leadership of the Canadian Wheat Board. And not just "someone" but, rather, someone with the knowledge, expertise and integrity to put philosophies and agenda aside and stick to the facts, as unbelievable as they are. Ken Motiuk has been a visionary for Canadian agriculture for over 40 years. His personal success as a farmer, and his legacy of service to the industry are testimony to who he is as a person. "Culture of Control" shares his experiences but, more importantly, offers a review of the history of the single desk selling, its pitfalls and its failings. Hindsight affords us the privilege of seeing the successful outcome of the removal of the single desk. This is the story of one man who held his ground for his beliefs and for the best outcome for western Canadian farmers. It is an important tome for future generations of farmers to read so they can understand the undertakings of those who built this industry. I highly recommend it for all generations."

Russ Crawford, author and grain industry veteran

≈

"In the book "Culture of Control", Ken Motiuk has done a terrific job in describing the history and impact of one of the most controversial institutions imposed upon the Canadian agriculture industry over the past 100 years – the rise and demise of the Canadian Wheat Board (CWB).

While there was some initial merit in the Government of Canada creating the CWB legislation to support and manage a fledgling Western Canadian grain industry during the war years of the early and mid 20th century, Motiuk describes in well-documented detail how the CWB outlived its usefulness, Eventually those good intentions became corrupted by a need for power, self-interest and politics to result in a highly dysfunctional grain handling and marketing system.

In meeting and getting to know Ken Motiuk during my more than 35-year career of observing and writing about the Western Canadian agricultural industry, from 1987 until today, I have always been inspired by farm men and women, like Motiuk, who for decades showed the commitment and courage to fight for their beliefs. Motiuk, not only a farmer, was often in the eye of the political storm, standing true to his convictions for the need for grain marketing freedom.

The photographs, illustrations and detailed description of the people and events in this book that eventually led to the official end of the CWB on Aug. 1, 2012, certainly took me for a walk down memory lane. While I was aware of the grain industry structure and farmer concerns, the book takes you behind the scenes, fills in the blanks, completes the picture.

In telling this story of the Canadian grain industry, Motiuk also weaves in the evolution and importance of improved technology and crop production practices, particularly over the last half century. If you ever wondered what events preceded construction of that network of 35,000 tonne capacity grain handling facilities across Western Canada, *"Culture of Control"* will be an informative and entertaining read."

Lee Hart, former field editor
Country Guide Magazine and Grainews

≈

"Culture of Control" is a textbook example of how much our agriculture industry is dependent upon good policy. On the same day that I finished reading this book, I met a young farmer who had never sold any wheat to the CWB. My mind flashed through the many faces and names of people who dedicated their time, sacrificed their own farms and families, and gravelled the road to grain marketing freedom. Many of those people are named in this book and all of them deserve thanks for the system we know today. I'm grateful to Ken Motiuk for sharing his account and I'm grateful to have this book as a gift to the next generation of farmers so they can be inspired to challenge the status quo. Read this book and learn what it takes to fight for change through the voices of farmers. As I see it…. we are facing a culture of control in many other aspects of our lives. We're gonna need more people like Ken to lead us."

Cherilyn Jolly-Nagel, farmer, Mossbank, Saskatchewan
Past president, Western Canadian Wheat Growers Association

≈

"I have known Ken as a friend and colleague for 40 plus years. I know, and when you read this book, you will know, Ken has devoted thousands of hours towards educating an industry and advocating selflessly for ALL Western Canadian farmers. He and his family should be proud of their accomplishments and the positive contribution Ken has made to all of agriculture on the Canadian prairies.

This is not only a story of the history of Western Canadian grain farming but, a story of a decades long battle to bring marketing freedom to Western Canadian grain farmers. It is a collection of facts that have framed the industry through nearly 150 years. The focal point of Ken's book accurately describes the trials of those who understood the impediments of "orderly marketing" and their relentless work to bring marketing options to a shackled industry. This is a good read for all who believe freedom is worth fighting for. Freedom is not free! As you will learn in reading this book, freedom can be lost at the stroke of a pen and it can take decades, if ever, to gain back."

Glenn Goertzen, Alberta farmer

≈

"By 1967 the trend of declining performance of the Western Canadian grain handling and transportation system was well established. By 1970, the marketing regime had deteriorated into its deepest state of decay. The situation had become so desperate that in the 1970-1971 crop year, grain delivery opportunities were based not on one's share of the wheat market, but on an area of land left idle, summer fallow. This practice later proved very detrimental to prairie soils.

This book offers the reader a detailed, firsthand account of the many insurmountable obstacles that farmers had to overcome in order to get to where we are today. There are few authors that could encapsulate the challenges faced by the advocates for change and modernization as well as Ken has.

This is the story of two tragedies. The first is that it took our industry 45 years to modernize itself into a reliable supplier to the world marketplace. The second is that many grain farmers and co-operative grain handlers resisted any form of change that would enhance our industry as a whole and take it to where we are today.

The generation that followed us has reaped the benefits of our work. They can deliver 40 tonnes or 400 tonnes to the elevator of their choice on any day they chose to meet their cash flow needs. Furthermore, they will never understand the torture 'that institution' put us through.

Ken has written a masterpiece that will serve as a factual account of historical events for generations to come."

Hubert Esquirol, farmer, Edam, Saskatchewan

≈

*"I grew up on a typical Saskatchewan grain farm, graduated from the University of Saskatchewan's College of Agriculture in 1980 and became immediately immersed in the reporting of agricultural news. Ken Motiuk's **"Culture of Control"** has the events, issues and people that made agricultural news through the intervening decades. Rather than just a historical account, Ken provides behind the scenes descriptions into the philosophical battles and clashing personalities.*

Although I don't share all of Ken's disdain for the Canadian Wheat Board, he provides his honest perspective as an industry insider and progressive grain farmer thereby bringing the history of grain production, transportation, marketing and politics into clear focus.

For those like me with grey/white hair, the book evokes thought-provoking nostalgia. For younger readers who have only heard stories of the CWB monopoly, quota acres, the Crow debate and the rise and fall of farmer-owned elevator companies, this book will allow them to connect the dots to better understand how agriculture in Western Canada evolved."

Kevin Hursh, agricultural journalist, agrologist and farmer

≈

*"Reading Ken Motiuk's book **"Culture of Control"** brought back many memories of how the Canadian Wheat Board affected the lives and careers of people other than farmers. I was raised on a mixed farm and in the 1960s I had to choose a career path. I recall my father having grain in the bins, enough to pay the bills, but unable to sell it because of the restrictive quota system of the Canadian Wheat Board. Being a farmer was out for me.*

After attending University, my wife and I purchased a feed mill in 1971. I soon learned the Canadian Wheat Board was still causing us issues. I had to purchase feed grains from producers who had no real idea of what their product was worth on the market. Initial payment – yes, final payment – who knows? It was a guessing game for both the producer and me. Some years if the final payment was substantial, the farmers I purchased from might have lost money. Other years my customers, the livestock producers, might have lost because I overpaid for the grain that was processed into feed.

Grain purchased by feed mills had to be entered into the Canadian Wheat Board issued permit books which always seemed to be kept at the elevator. Another stop for the producer to pick his book up and explain why he needed it. Now if he happened to have an advance from the Wheat Board, I would have to deduct a set amount per bushel off his cheque and submit it to the Wheat Board much to the chagrin of the producer. I wasn't done yet. I was subject to regular inspections by the local representative of the Wheat Board to ensure I was in compliance. Very time consuming for us.

Ken's book is a well documented history of the Canadian Wheat Board. A must read for anyone associated with the agricultural industry."

Glenn Keddie, past owner of
Grande Prairie Feed Service (1971) Ltd.

≈

"I've known Ken for many years and always known him to be an involved advocate for positive change and evolution. Thankfully he has shared his experiences and insights into the industry about which he clearly feels passionate, adding context and background to the events of which we were once aware, but now, also well-informed. In **"Culture of Control"**, *Ken gives an insider's view into the past archaic workings of THE most important industry – food production – and the culture of control that permeated it for far too long. Two thumbs up!"*

John De Pape, commodity trader and risk advisor

A hazard of getting old is that one may be smitten by the urge to write a personal or family history. It is debatable whether this affliction arises from a desire to inform present and future generations about the life and times of the author or is merely an ego trip based on a belief that the experiences of one's life were significant enough to merit recording. Let the cause be what it may; the onset of this affliction can be regarded as a sure sign that the victim has reached old age. Having qualified in that respect, and succumbed to the malady, my course will now be along the well-trodden path of countless previous sufferers.

– Mac Runciman, President United Grain Growers, 1961–1981[1]

---

1   Earl, Paul D. *Mac Runciman, A Life in the Grain Trade.* University of Manitoba Press. 2000. Page 3.

PREFACE

During the reclusive period of the COVID pandemic, I was in our farm shop one day and Justin Leliuk, our son-in-law who manages our family farm, asked about the Canadian Wheat Board (CWB) and how it worked. Three of our young employees were also there, so I started explaining what it was like to farm with the CWB in control of grain marketing.

As I began to explain, I realized there was a story to tell, as most young farmers have no idea what it was like. I decided I had to tell the story from my perspective. Since I had both farmed and been involved in many farm organizations, I had a personal, unique, and multidimensional perspective of the events that transpired over the four-decade-long deregulatory transformation of the Western Canadian grain-marketing-and-transportation system.

Wendy and I started farming in the 1970s. At this time, more and more farmers in the West were becoming disillusioned with the dysfunctional system that had emerged and the paternalistic attitude of the CWB and the Prairie Wheat Pools. The voices for change were becoming louder and farm organizations more sophisticated and knowledgeable in their arguments for reform. I was but one cog in a wheel that was turning faster and faster. The success of everyone's combined efforts eventually achieved the results we were looking for—a modern, deregulated, and market-responsive grain-marketing-and-transportation system.

For those who have never had the experience of selling grain through the CWB, here is my story.

Ken Motiuk

Those who do not learn history are doomed to repeat it.

Philosopher George Santayana

This narrative is an account of the evolution of the Canadian system for marketing, handling, and transporting prairie grain over the past 130 years.

The Canadian prairies were settled in the late 1800s. Virgin soil was broken and sown to wheat. The first thirty years witnessed spectacular growth as immigrants, mainly from Europe, settled the vast lands. At this time the infrastructure of the wheat-export industry was controlled by Winnipeg-based grain companies and the Canadian Pacific Railway (CPR). There were widespread complaints from the farm community about abuse of market power and unscrupulous behaviour by the grain companies and the CPR. Repeated requests came from the farm community to have the government intercede with regulations. This became a precondition to the regulatory era that accelerated during World War I.

Over the next fifty years, from World War I through to the 1970s, a sequence of events transpired that culminated in control of the entire system being vested in the hands of the Canadian Wheat Board (CWB), a federal government agency. These events included government control of wheat supply during both World Wars, the rise of the farmer Wheat Pool movement in the 1920s, the hardships of the great depression, and the execution of worldwide wheat-trade agreements on behalf of the federal government in the post–World War II period. Grain companies, railroads, and farmers lost complete control over the grain (primarily wheat) marketing, handling, and transportation systems.

By the 1970s, the system was in total disrepair as the railroads and grain companies had stopped investing capital in assets they had no control over. The regulatory bureaucracy, led by the CWB, was entrenched and interested primarily in self-preservation. In 1972, the Soviet Grain Robbery resulted in an explosion in world grain prices and demand for grain. Prairie farmers responded by growing more grain. The outdated and hyper-regulated prairie grain marketing, handling, and transportation system could not get grain from prairie points to tidewater for export. Millions of dollars of grain sales were lost. This was the precondition for the next fifty years that witnessed the reversal of the regulated system to one based on market principles.

Meanwhile, in the 1970s, we were trying to grow our prairie farm. World grain prices were at record highs and trade was booming. Yet in Western Canada we had a backlog of grain and nowhere to sell it. The following is an account of my personal involvement in the deregulatory phase and transformation of the systems of marketing, handling, and transporting Prairie grain over the past forty-five years.

Nobody is qualified to become a statesman who is entirely ignorant of the problem of wheat.

–Socrates

INTRODUCTION

Wednesday, November 22, 2006, was the date of a memorable meeting at 423 Main Street in Winnipeg, Manitoba, the head office of the Canadian Wheat Board (CWB). Greg Kelly, a Saskatchewan farmer and chairman of the board of Prairie West Terminal at Dodsland, Saskatchewan, along with the terminal's general manager, Garth Gish, travelled to Winnipeg to meet with CWB officials. They represented the owners of a new high-throughput grain terminal built by local farmers. The topic of discussion at the meeting was Prairie West's inability to obtain sufficient railcars from the CWB to ship grain that Prairie West was purchasing from its farmer owners. Greg set the tone of the meeting in his opening remarks: "I feel quite intimidated here. What you might not realize is that you (the CWB) control where I sell my crop. You control when I sell my crop. You control how much of my crop I can sell. You control the price I get for my crop. You control my entire livelihood."

Greg's comments reflected the views of many Western Canadian farmers, independent business operators constrained by a yoke of bureaucratic control, frustrated by restrictive government regulation, and fed up with selective regional legislation that stifled the growth and prosperity of one of Canada's greatest renewable natural resources.

The production, consumption, and trade of wheat has been of paramount importance in the history of humanity. Wheat is arguably one of the most important crops grown by mankind, rivaled only by rice. It is the most widely grown cereal crop and a staple food for 35% of the world's population.

Wheat is an ancient grain that was domesticated about 12,000 years ago. When milled into flour, wheat is the main ingredient of bread, a staple in the human diet. Wheat flour is also used to make noodles, flatbreads, cakes, and cookies. When the Canadian prairies opened to settlement in the late nineteenth century, settlers found this area to be well-suited to the production of wheat. Canada very soon became a major player in the world wheat market.

In 130 years, the Western Canadian system of handling, marketing, and transporting Prairie grain has come full-circle in the approach taken to managing the system. The evolution began with a free-market approach in the late nineteenth century, through to a rigidly regulated system from the post–WWII years through to the 1970s, and then back to a market-driven system by the early 2000s. In my various roles, primarily as a grain-producing farmer, I experienced an often-frustrating journey, participating in and encouraging the deregulatory element of that evolution.

My involvement started in 1979 when Don Mazankowski became minister of transport and minister responsible for the Canadian Wheat Board (CWB) in the short-lived Joe Clark Conservative government. It continued through to the dissolution of the CWB in 2015 under the Harper Conservative government. Being involved was a tremendous learning experience, with many highs and lows. Ultimately it was very rewarding. I was able to meet and work with many interesting people. Progress was slow, but those of us determined to see positive change persisted. In the end we achieved what we set out to accomplish, a market-based grain-marketing-and-transportation system capable of meeting the demands of the twenty-first century. It was a privilege to have been part of it.

There were essentially five components involved in this process: farmers, grain companies, railroads, the federal government, and the Canadian Wheat Board. Over a forty-year period I worked in four of these entities and was able to see the problematic situation from the cross-perspective of participants. Throughout this time, I wore many hats: a farmer producing grain, an employee of a federal minister's office, and later a sitting member of numerous boards. I worked at the Grain Transportation Authority; I was a director on the board of United Grain Growers; and lastly, I was a director of the Canadian Wheat Board. These positions helped me see the situation from a total-system perspective by observing the inner workings and interactions of the various parties.

GROWING UP ON THE FARM—THE '60S AND '70S

I graduated from the University of Alberta in 1976 with an agriculture degree and started working with Alberta Agriculture. On evenings and weekends, I was back at the farm in Mundare, Alberta, helping my father, since farming and tractors have always been a passion for me. The problem was that the farm never seemed to make much money. We had a small, mixed farm (growing field crops and raising livestock), and it was managed in a very conservative and traditional fashion.

Mike and Adelaide Motiuk farm, 1969

Our small, mixed farm at Mundare was common for the district and the era. This part of northeastern Alberta was settled by Ukrainian immigrants around the turn of the nineteenth century and that is my heritage. On our farm, we had some beef cows that were pastured on non-arable land. Calves were fed over the winter and sold in spring as feeders to be fattened in a feedlot before slaughter. We had a small number of dairy cows, milked by hand, with the cream being sold for household income. A small number of laying hens were kept, their eggs being sold for living expenses. Broiler chickens were raised every summer and then butchered and sold privately, again with proceeds used for domestic expenses. At times, we had a small number of hogs that were also sold for slaughter.

The main enterprise on the farm was grain farming. In the late '60s we farmed a half-rented, half-owned section of land. Total arable acreage was about 450 acres. Our farm was of average size for the district, and as was common in the area at the time, a three-year crop rotation was followed. Land left to fallow the previous year was sown to wheat. Oats were sown on the wheat stubble. The third year the land was left fallow again to cultivate and reduce the weed population and "rest the land." This crop rotation was common in our area. Wheat was sold as a cash crop, and oats were used as feed for livestock.

Our land base was primarily rich and fertile black chernozem that had attracted the Ukrainian settlers to this area ("cherno zemla" means "black soil" in Ukrainian). The only other place in the world where there is chernozemic soil is in Ukraine. This was the parkland belt of the prairies and much of the land had been heavily treed when it was settled. The early surveyors called this area the fertile belt, as the soil was richer and the rainfall more plentiful than in the southern and central grassland prairies. However, by the 1960s the natural fertility of the soil was diminishing, and the land was heavily infested with wild oats.

During this time that I was growing up on the farm, the system for marketing, handling, and transporting grain in Western Canada was heavily regulated by the federal government. There were very restrictive measures in place to sell prairie grains and oilseeds. The federally mandated Canadian Wheat Board (CWB) had a monopoly on wheat, barley, and oat exports from Western Canada. Through a tightly controlled delivery-quota system, prairie farmers were bound by law to sell their wheat, oats, and barley to the CWB. The CWB also had the authority to set delivery quotas for the so-called non-board crops, which included rapeseed (later called canola), flaxseed, and rye grown in the West. These non-board crops were marketed directly through the grain companies.

The CWB had jurisdiction over the marketing of farmers' wheat, oats, and barley in Alberta, Saskatchewan, Manitoba, and the Peace River district of northeastern British Columbia. This area was called "the Designated Area." Farmers in Ontario, Quebec, Atlantic Canada, and British Columbia other than the Peace River district were not under the jurisdiction of the CWB and could sell these grains directly to grain companies without quotas.

The CWB did not own any grain elevators, so they paid handling and storage fees to grain companies for the wheat, barley, and oats purchased from farmers. Small towns and railroad sidings across the prairies had anywhere from one to eight small elevators where farmers delivered their grain. The grain elevators were owned by grain companies such as United Grain Growers, Pioneer Grain, National Grain, Searle Grain, Federal Grain, Alberta Pacific Grain, and the three Prairie Wheat Pools. These companies simply warehoused CWB grains, while they were merchants of non-board grains including rapeseed (canola), flaxseed, and rye. Rapeseed (canola) and flaxseed are more appropriately termed oilseeds, but for the purposes of this narrative all these six crops will be referred to as grain.

These six main crops grown on the prairies—both board and non-board—were called the six major grains. Wheat, barley, and oats were marketed solely by the CWB (board grains), and rapeseed, flax, and rye (non-board grains) were marketed directly by the grain companies. The CWB controlled individual farmer deliveries of all grain through the delivery-quota system, and it also controlled the allocation of railroad-owned grain cars to ship these products from country elevators to final destinations. These destinations could be Canadian domestic processing facilities such as flour mills and malt plants. Of much greater volume and importance were shipments to tidewater for export to overseas markets. Export grain was shipped either westward to Vancouver, or eastward to Thunder Bay for trans-shipment to terminals on the St. Lawrence River where it could be loaded onto ocean-going vessels.

The CWB negotiated the railcar supply for grain shipment with the railroads, and then decided at what country elevator the cars would be spotted (positioned for loading), and which product was to go into them. The CWB also directed at which terminal facility at Thunder Bay or Vancouver each railcar was to be unloaded.

Through the late 1960s the CWB was not aggressive in selling wheat, and Canadian stocks rose while world prices fell. Since farmers and the private grain trade were not able to sell wheat, oats, and barley to anyone other than the CWB, farm income depended on how well the government-mandated CWB monopoly did its job.

≈

Working at Alberta Agriculture, as well as meeting fellow aggies from all parts of the province at university, I became aware of alternative and more progressive ways of farming. More advanced farmers in the province were using nitrogen fertilizer and spraying wild oats out of domestic crops with new herbicides. More progressive farmers were adopting continuous cropping, that is, ending the practice of summer fallowing or "resting the land," as traditionalists called it. Eliminating summer fallow gave farmers a crop every year from their land, not just two years out of three. Progressive farmers were also starting to grow new crops such as canola on their farms.

My responsibilities at Alberta Agriculture were not that demanding, and with the encouragement of my boss, Phil Jensen, I researched how we could adopt these new practices on our farm. They were much more input-intensive, so more cash outlay was required, and thus more risk. After much discussion, cajoling, and no small amount of disagreement, I convinced my very conservative parents to begin adopting some of these progressive farming practices on our farm. The small livestock enterprises on our farm were phased out, and we focused on grain farming.

On our farm in 1977 we seeded canola, ended the practice of summer fallowing, and began using anhydrous ammonia nitrogen fertilizer and wild oat herbicides. The rich black soil responded well and we produced good crops. Cash expenses were higher, but grain prices were also high, so the economics of the budgets worked. After a couple of years of increased production, we found we were unable to sell all the grain we were growing. The CWB delivery-quota system was restricting both *what* we could sell and *how much* we could sell, so we couldn't empty our bins annually to pay the extra expenses. I began researching this more because I could not understand why, despite the world price of, and demand for, grain being high, we could not sell our grain grown in Western Canada.

Country elevators were plugged with grain. Ships were waiting for grain in Vancouver, but we could not deliver from the farm. Demurrage (penalty paid by a shipper for not delivering a product by a contractual date) was being paid on these waiting vessels. Canada's reputation as a reliable supplier of grain was diminishing. We were losing out on hundreds of millions of dollars of lost grain sales. It was evident the entire grain-marketing-and-transportation system in Western Canada was not working.

The CWB oversaw the logistics of grain movement. They made the sales and allocated available railway boxcar supply to the grain companies based on historical handling percentages. If a company historically handled 15% of the grain in the designated area, they would get 15% of the railway boxcar allocation from the CWB. Though farmers could choose which grain company to deliver their CWB grain to, this option was not always available if the CWB did not allocate railcars to the elevator an individual chose to haul his grain to.

Grain companies had no way to increase individual company market share in this system. If the CWB did not open a delivery quota because of lack of sales or shipping problems, farmers could not deliver grain. Or the quota could be opened but the local grain elevator could not accept additional grain deliveries from farmers since there was no space, as the CWB had not allocated cars to that facility.

The CWB blamed the railways for not supplying enough boxcars for grain movement. The railways did not always have boxcars available and were not eager to invest in new equipment and pull grain to export position at tidewater since they were losing money hauling grain due to the Crow's Nest Pass Agreement. The system was gridlocked and dysfunctional due to regulatory obstruction. Farmers on the prairies watched helplessly as great revenue opportunities were lost with every failed grain sale. Added to this were labor disruptions in Vancouver as unions were demanding higher wages to address the high inflation of the 1970s.

So why was everything so messed up?

The problem was too much regulation, too many regulatory bodies with overlapping jurisdiction controlling the system, lack of market signals to farmers, and a disinterest from railways in hauling grain because they were losing money on every bushel they shipped.

This monopolistic structure was supported by self-serving farm organizations, led by the Prairie Wheat Pools, who continually planted fear and uncertainty in the minds of farmers that any change to a market-driven system would be to the demise of their finances and the prairie farm economy. How did we manage to get to this dysfunctional point?

There were many "layers to this onion"—there was no "silver bullet." The entities (grain companies and railroads) that had invested money and built and owned the assets in the logistics chain had no control over how product flowed through their assets. This was all in the hands of the CWB, which controlled sales, delivery quotas, and railcar supply to grain elevators. We had a government agency (the CWB), run by government-appointed bureaucrats, decision-makers with no financial interest in the system, controlling the assets and supply chain. The CWB's command-and-control system did not allow for the railways, grain companies, or farmers to seek innovative solutions to emerging problems. This resulted in a major disincentive for the railroads and grain companies to invest in capital improvements in the system.

IN THE BEGINNING – BUILDING THE RAILROAD

To fully understand how we arrived at this point in the 1970s, we must look at history to see how this all evolved.

In 1867 there were four provinces that were part of the original Canadian Confederation: Ontario, Quebec, Nova Scotia, and New Brunswick. The land from the Canadian Shield in northern Ontario, across the prairies, and over the Rocky Mountains was populated only by small groups of Indigenous peoples, missionaries, and fur traders. On the Pacific coast there were small British colonies on Vancouver Island and the lower mainland. In 1871 Prime Minister Sir John A. MacDonald created the Province of British Columbia with a promise to build a railway to connect the vast distance between Upper Canada and the Pacific Coast.

The MacDonald government was becoming worried that the Americans had their eye on this western territory. Minnesota was looking northward along the Red River Valley towards Winnipeg, the fertile land north of the 49th parallel and the site of the fledging Selkirk settlement. American ranchers from Wyoming and Montana were pushing northward along the eastern slopes of the Rocky Mountains into what is now southern Alberta to pasture their cattle. Along the West Coast, Americans were also gazing northward towards what is now Burrard Inlet and the adjoining land where the city of Vancouver is now located.

To fulfill the promise, the government urged a group of Montreal businessmen to form the Canadian Pacific Railroad (CPR), which undertook to build a transcontinental railroad. This railway from southern Ontario to Vancouver was completed in 1885. Since the government had no funds to pay the CPR, the railroad was granted vast acreages of unsettled land on the prairies[2] as well as rights to mineral concessions in the mountains. The CPR then had to find a way to turn these lands and concessions into cash.[3] Now both the government and the CPR had a common interest in having people come and settle on the prairies. Both the federal government and the CPR sent land agents to Europe to entice people to emigrate to Canada, settle on prairie land, and become independent farmers. Many of these settlers were from rural Ukraine, where the Austro-Hungarian regime of oppression was obstructing their freedoms and economic livelihood.

In 1897 the government wanted a railway built through the Crowsnest Pass and into the Kootenay area, where there were coal deposits in the mountains. Once again this was to forestall encroachment of the Americans into this region. To entice the CPR to do this, federal land and mineral grants were once again given to the CPR to pay for construction costs. In return, the CPR agreed to ship grain from the prairies at fixed rates into the future—the origin of the Crowsnest Pass Rates. By the 1970s, due to inflation and increasing costs, huge losses were being incurred by the railways hauling grain at the fixed 1897 rates.

---

2   This was the same concept employed to finance the construction of the American railroad system between 1850 and 1870.

3   Berton, Pierre. *The Great Railway*. McClelland & Stewart Inc. 1974 is an excellent account of the construction of the transcontinental Canadian Pacific Railway. It outlines all the political scandals, bribery, and corruption that occurred in the letting of contracts for the survey and construction.

THE EMERGENCE OF THE WHEAT ECONOMY

It was in the best interests of both the Government of Canada and the CPR to have the West settled and the land broken for wheat production, for which the Canadian prairies were naturally suited. This created revenue for the CPR, as they sold the land they had been granted to settlers and then transported the settlers and their effects to the prairies. The Government of Canada paid for CPR transport of immigrant settlers by rail from ports in Halifax or Quebec City where they had landed, to their prairie settlement locations. Once established, these settlers grew wheat, which the railways shipped to market. As well, manufactured goods produced in Eastern Canada were purchased by the settlers and freighted to the prairies by the CPR. This created long-term financial sustainability for the railroad.

Western settlers were forced to pay a stiff tariff on goods such as farm equipment that were purchased from Eastern manufacturers. The same goods could be purchased cheaper from the US, but the tariffs protected against their importation at competitive prices. This policy structure was as much a plan to provide for economic growth by the manufacturing sector in Eastern Canada as it was to settle and develop the Prairies. Eastern Canadian manufacturers profited, the CPR profited, and the Government of Canada profited. Homesteading settlers, mainly from northern, central, and eastern Europe learned of the vast potential of the area and felt fortunate to become landowners. They were lured by the Homestead Act, whereby they could purchase 160 acres of land from the government for $10. Settlers could add to their homesteaded landholding by purchasing additional land from the CPR. It was by these metrics that the Prairies were settled. All the while, the growing number of settlers in the West was enriching the tariff-protected industries of Central Canada.

The tariff on manufactured goods from Central Canada was one of many federal policies that favoured Ontario and Quebec at the expense of the western provinces.

In 1876 the first shipment of wheat left Winnipeg moving southward up the Red River by steamer to Fargo, across to Duluth by rail, on to Sarnia by laker, and then by rail to Toronto. Later, the newly constructed CPR railway connected Winnipeg to Port Arthur/Fort William on the northwest end of Lake Superior, now known as Thunder Bay or The Lakehead. There the wheat was loaded on a lake vessel and shipped to Ontario. Wheat could now move from Winnipeg to Ontario, utilizing an all-Canadian route.

As settlement progressed, Winnipeg became the "Gateway to the West." All new immigrants passed through Winnipeg on their way to their new homesteads in Manitoba, Saskatchewan, and Alberta. The grain companies and Winnipeg Grain Exchange were located here. The western base of the CPR, with rail yards and shipping and service facilities, was located here, as well as a light-manufacturing sector serving the western provinces. It was the financial and insurance centre for early prairie settlement. For the next one hundred years the production and export of wheat dominated Western Canada.

Settlers kept coming to the West. Small grain elevators, dominated by business interests from Winnipeg, were built on sidings provided by the CPR. These sidings grew into prairie towns. Since there was no other railroad, the CPR had a monopoly on transportation. There was an unwritten alliance between the CPR and Winnipeg grain companies. The grain companies that built elevators along the rail lines became known as "line elevator companies." Milling interests such as Ogilvie were also building elevators to gather and ship wheat to their eastern mills.

Before long, farmers became upset over unfair treatment by the CPR and grain companies. There were issues regarding railcar supply; grading disparities; high dockage and short weighing; high handling tariffs; and essentially unscrupulous and dishonest treatment of farmers who were captive to the CPR/grain company combine. Farmers attempted to order and load their own railway boxcars for shipment of wheat, but the CPR favoured spotting railcars only at the grain company facilities.

Frustrated with the CPR/grain company combine, eventually these early settlers obtained the attention of the politicians, and in 1900 the Manitoba Grain Act was passed, the forerunner to the Canada Grain Act and the current Canadian Grain Commission. The Act established standards for official weights, dockage, and quality grades for wheat. But it did not solve the problem of the CPR still refusing to allocate boxcars directly to farmers to "load over the platform" (later known as producer cars). This was finally challenged in the courts. The historic decision that occurred around events at Sintaluta, Saskatchewan, in 1903 directed the CPR to establish a public car order book and supply boxcars directly to farmers who wished to bypass the grain company elevators.

The dissatisfaction with the CPR and grain companies by the farm community continued. Some farmers decided to band together and start handling and marketing their own grain. In 1906, the Grain Growers' Grain Company was established by a group of business-minded Manitoba farmers. Soon after that, the Alberta Co-operative Grain Company was established, along with the Saskatchewan Co-operative Elevator Company. In 1917, the Grain Growers' Grain Company and Alberta Co-operative Grain Company amalgamated to form United Grain Growers Ltd. The Saskatchewan group chose to stay out of this merger and remained independent, later becoming part of the roots of the Saskatchewan Wheat Pool.

The West was booming as settlement and industry grew through most of the first two decades of the twentieth century. Wheat production and export records were repeatedly set and subsequently broken and then set again in this period. Central Canada wanted the high-quality wheat from the west, and internationally the reputation for quality Canadian wheat was growing. The prairie grain elevator and the row of elevators along a railroad siding became international symbols of bountiful wheat production. The Canadian Prairies were now a major supplier of wheat in the world market.

The CPR was all-pervasive at this time. It provided land on which the grain companies built their elevators. The CPR owned the land around these sidings, and small towns evolved. CPR land was sold to businessmen who provided goods and services to the farmers settling in the community. The CPR brought the settlers and their effects into the area. Then goods such as coal, lumber, twine, fencing supplies, farm equipment, and other manufactured goods required by the settlers were brought in by rail. This also included household goods and supplies, and later, household appliances. You could even order a house from the Eaton's catalogue, and the CPR would ship it to your town. The grain companies constructed warehouses along the rail line to receive shipments of these goods. Elevated, flat, earthen platforms with gradients sloping to ground level were constructed along the rail lines to unload farm equipment from railcars.

In the 1950s, when farmers began to use herbicides and fertilizers, grain company warehouses stored bagged fertilizer and five-gallon pails of herbicides for sale to farmers. Fertilizer was shipped in eighty-pound bags, which had to be manually carried from the rail boxcar to the grain company warehouse. As a young boy, I remember watching bags of fertilizer being unloaded in this fashion.

After bringing goods into the community, the CPR would ship wheat from the line elevators to the large grain terminals at the Lakehead (Fort William and Port Arthur, Ontario, on the western shore of Lake Superior). Later, as livestock production increased, livestock were shipped to market from the collection yards built along the rail line. At this time, most of the grain terminals at the

Lakehead were owned by the CPR. Grain elevators and terminals were all painted "CPR red" (a rusty reddish-brown), no matter who owned them. This colouration dominated until the 1960s when some grain companies started using their own corporate colours and images. Indeed, the CPR was the lifeline to all prairie towns at this time since a network of highways and roadways had not yet been built.

The CPR, which dominated all Prairie life, was not always favoured by farmers. This is well illustrated by the illogical tale often told in the farm community about the farmer who had just lost his crop to a hailstorm and looked upward at the sky, shaking his fist and saying, "Goddamn CPR."

Until the 1920s, the CPR was the only overland artery that connected Ontario and the West Coast. It totally dominated development of the city and the Port of Vancouver at this time. By mid-century, the CPR was a huge conglomerate with ocean-going vessels that transported passengers and others that transported freight. It owned a chain of hotels and was involved in tourism. It operated the second largest airline in Canada. It had significant assets involved in mining on the lands granted to it by the federal government when the first transcontinental line was being built.

The Western wheat economy was funding the construction of a large grain-handling-and-transportation infrastructure. To provide some competition to the CPR, two new railways, the Canadian Northern and the Grand Trunk Pacific, began to build two new transcontinental lines across the northern prairies in the early 1900s. Grain companies kept building elevators on new railway branch line sidings on the prairies as well as large transshipment terminals at Thunder Bay (as Fort William and Port Author became known) on Lake Superior.

Small towns were springing up around the railway sidings and grain elevators that dotted the Prairies. A shipping industry for moving wheat on the Great Lakes was growing. The Winnipeg Grain Exchange became the world leader in setting the international price of hard red spring wheat. Things were becoming favourable for wheat production and growth of the Western economy. Britain was the largest buyer of wheat in the world and Canada was the largest supplier. The Port of Montreal was the largest port in North America.

Steam train pulling into Mundare, Alberta station – 1920s
Courtesy of Prairie Towns

THE FEDERAL GOVERNMENT BUILDS
LARGE GRAIN TERMINALS

Even though farmers had organized to be able to market and handle their own wheat, they were still displeased with the large grain companies and the CPR. These groups of farmers, such as the Grain Growers' Grain Company, did not have the capital to build elevators. Established grain companies and the CPR were not friendly to their motive of shipping their own wheat. These farmers had difficulty negotiating fair and accessible handling agreements with the established line elevator companies and the CPR to handle their grain in a timely manner at a reasonable cost. A strong lobby from the farm community wanted the Canadian government to build grain-handling facilities for use by the farmer-owned grain companies. Farmers wanted these terminals to be declared a public utility so they could be used by anyone.

The line elevators that were built on country rail sidings were usually smaller and of wooden construction. They were designed to receive grain from farmers who hauled their wheat from the farm by horse and wagon. Here the grain was weighed, inspected, and assigned a grade, and then it was elevated and dropped into bins in the facility. When railcars came, the grain was elevated again and then dropped into railcars through long discharge pipes. The wheat was then shipped to the Lakehead terminals where the grain was cleaned and stored for shipment to final market. These port terminals were much larger than country line elevators.

Between 1913 and 1916, the first Canadian government grain terminals were built. Inland facilities were built in Saskatoon, Moose Jaw, and Calgary. Port terminals were built at Port Arthur on the Lakehead and the newly expanding Port of Vancouver. These were relatively large structures and the first in the West to utilize concrete slip form construction.

Dominion Government Elevator, Moose Jaw, Sk, 1917
Industrial Building in the West. Photo credit: Patricia Vervoost, Saskatchewan Archives Board #R-B9680

An inland terminal of similar size and design was completed in Edmonton in 1924 and another in Lethbridge in 1931. Port terminals were completed in Prince Rupert in 1925, and after a long-drawn-out process and many delays, the Churchill terminal was completed in 1931.[4]

The Churchill facility, plagued with natural hurdles, was never used to full capacity and has essentially been closed since 2024. The shipping season out of Churchill is only about three months a year due to ice conditions in this sub-Arctic location. The more northerly section of the line is built on muskeg, and the permafrost base upon which the rail line rests is not stable. Trains must go very slowly while pulling smaller railcars that are not fully loaded. This is costly from a rail perspective. Also, insurance rates on vessels serving Churchill are high due to the possibility of collision with an iceberg.

All these concrete terminals were futuristic in design at the time, and the inland terminals were considerably ahead of their time in both form and function. They were large facilities that could clean, store, and ship grain in unit trains[5] ready to load to an export vessel. Geographically, they were located a long way from prairie farms, and it was not practical or economical to haul grain there with horse and wagon box. Later, when trucks were hauling grain longer distances, these terminals were never used to full capacity until the freight rate structure was changed in 1996.

Likely the main reason these large facilities were under-utilized for such a long time was the Crow Rate structure. Rail charges were fixed at 1897 levels and were all strictly single-car and distance-related. The rail freight rate at any shipping point was calculated by distance to port, with no consideration for whether the point was on a heavy-traffic main line, which was much easier and less costly to serve, or a little-used branch line. Additionally, the distance-based freight rate did not reflect a more efficient larger car spot at a newer elevator facility versus a one- or two-car spot at an old, crumbling facility.

---

4   The story of the Port of Churchill is well told by McEwan, Grant. *The Battle for the Bay, The Story of the Hudson Bay Railroad*. Western Producer Book Service. 1975.

5   A unit train is a full train of cars all loaded with the same commodity and pulled as a "unit" to port position.

Variable rates—a varying rate structure that reflected more efficient handling and loading facilities—were not allowed under this federal legislation. Thus, a lower rate structure reflecting railways' efficiencies at a large car spot was not possible. Later, in the battle to change the rail rate structure in the 1980s, the Wheat Pools were the strongest supporters of the existing Crow Rate. They had over 60% of the small grain elevators and the fixed single-car rate served them well with their small-car spots. The Pools did not wish to see a modified rate structure that would reflect rail efficiencies and likely direct more grain to the inland terminals with their large car spots. All this changed with new railway legislation in 1996.

The Government of Canada owned these under-utilized facilities until they were sold in 1979. The Alberta government purchased the three Alberta facilities and operated them as Alberta Terminals Ltd. They were sold to Cargill in 1991, who started to put them to good use once discounts for multiple railcar spots were allowed. Cargill closed the Edmonton facility in 2024, one hundred years after it was built. The Calgary facility has been demolished.

The Saskatoon terminal was sold to Northern Sales in 1979, and the Moose Jaw terminal was sold to Allstate Grain in 1981. The Saskatchewan terminals were later purchased by AgPro Grain (a subsidiary of Saskatchewan Wheat Pool). Over time the port terminals at Vancouver and Thunder Bay were sold to the mainstream grain companies.

The Prince Rupert facility was sold to a consortium of the six main grain companies in 1980.

WORLD WAR I, THE FIRST WHEAT BOARD, AND THE ESTABLISHMENT OF THE PRAIRIE POOLS

"Wheat pools have become as much a religious institution as the church."

–Henry Wise Wood, Founding President, Alberta Wheat Pool

With the onset of World War I, production, consumption, distribution, and pricing of wheat worldwide were disrupted. The British were at war and were worried about a secure supply of wheat. At this time, Canada was still a British colony, and the British instructed the Canadian government to take control of wheat supplies. Using power vested in the War Measures Act, in 1917 the federal government took control of wheat price, supply, and marketing in *all* of Canada. This measure served to prevent hoarding and profiteering during this period of international conflict and uncertainty. The intent was clearly to oversee the supply of wheat and prevent the price from skyrocketing while ensuring a secure supply of wheat for the British. These objectives were later administered by the formation of the first wheat board in 1919, which was meant to be a temporary measure until the war ended and markets stabilized.

Short world stocks due to production disruption in Europe caused wheat prices to go up during the war. Adding to this, short crops in Canada in 1918 and 1919 tightened stocks even further. Gradually, wheat production, consumption, and trade normalized in the early 1920s. Stocks of wheat were restored, production and trade stabilized, and the price of wheat fell. This wheat board was disbanded in 1920 as the War Measures Act was rescinded.

Because this sequence of events occurred at the same time, many farmers erroneously believed that it was the creation of a wheat board that had resulted in prices going up during the war. However, the reason prices went up was the high demand from Great Britain as well as two short crops in Canada. This first wheat board had little role in setting these higher prices, and in fact, the regulations it employed likely played a role in maintaining a limit on wheat prices. This assisted the war effort since it provided price stability for wheat consumers in this turbulent time of conflict.

As wheat prices drifted downward during the postwar depression of the early 1920s, the appetite for many farmers in the West to take control over wheat marketing grew. Farmers were frustrated with low grain prices and there was ongoing dissatisfaction with the line elevator companies. They began banding together in co-operatives to establish the Prairie Wheat Pools, the intent of which was pooled marketing of wheat, much like the first wheat board had done during WWI. Their ultimate objective was to have all Canadian wheat sold through one big pool and be able to control the world price. The idea grew as quickly as a prairie fire. In short order, leaders of the movement worked successfully to convince farmers to contract their wheat to a pool. Prairie farmers soon contracted 50% of their wheat crop to the pool.

Each Pool was given a charter to operate in its respective province. The Alberta Wheat Pool (the AWP) was chartered in 1923, and the Saskatchewan Wheat Pool (the SWP) and Manitoba Pool Elevators (MPE) in 1924. While United Grain Growers (UGG) operated under federal legislation, the Prairie Pools operated under provincial legislation.

UGG was a farmer-owned grain company, a business venture operating as a co-operative. The Pools were more of a political, socioeconomic movement. Not only did they handle farmers' wheat through a pool, but they also acted as a farm organization speaking on behalf of Western farmers. Their political lobby became strong and influential. The entire co-op movement had elements of a change in social structure for prairie agriculture as farmers embraced cooperation and banded together to do business as co-operatives. Pools and co-ops were also a change to the economic structure as private enterprise and profit were being replaced by co-operatives and member sharing of profits.

The Pools were rather parochial organizations. Their strength was anchored in their unwavering belief that the co-operative structure would become the new economic order. It would replace the existing laissez-faire, unregulated free market, which they felt had failed them. Co-ops and the pooling of some agricultural product sales was starting in various US markets and other places in Canada. But nothing was as large as the wheat pool movement in Western Canada.

UGG differed from the Pools in their underlying beliefs. Although they were structured as a co-operative, they believed the marketplace, not government intervention, should guide the evolution of the system. They strongly believed in farmers' freedom to run their own businesses. UGG's farm policy position always supported market-based initiatives.

The Pools were quite leftist in their policies, supporting more regulatory efforts by government, the most doctrinaire being the SWP. On issues such as grain marketing and the Canadian Wheat Board; maintaining the Crow Rate; and overall government involvement in agriculture, the Pools and UGG differed substantially. The Pools later had a stronger voice in agricultural policy, citing their large membership of farmers. It was a great irritation to the Pools that UGG could say it represented farmers as well and had an opposing view on policy solutions.

UGG's equity structure was made up of farmer-owned shares, with dividends being paid on these shares when the company was profitable. The Pools distributed company earnings (patronage dividends) in proportion to each member's business with the company. The Pools closely followed the political philosophy of "one man, one vote." It was the objective of both the Pools and UGG to eliminate the middleman and return all profits to farmers.

Many farmers belonged to both a Pool and UGG, which had totally opposite positions on farm policy. If you added up all the members of the Pools and UGG, the sum was significantly higher than the number of farmers on the Prairies, yet each entity purported to represent its members' views on farm policy.

These four entities were all co-operatives, with the description being "working together for a common cause." This is clearly spelled out in the Rochdale Principles. As the Pools became more powerful, they challenged UGG as not being a real co-op since profits were distributed through shareholdings rather than patronage. They would attempt to discredit UGG's market-based policy perspectives since this was in direct opposition to the regulatory approach advocated by the Pools. In this manner, the Pools claimed that only *they* were the real voice of Prairie farmers.

Western farm politics became divisive at this point. The entire pool movement was forceful in attempting to convince farmers to support them. Farmers who chose to remain independent of the pool movement were often coerced and cajoled into joining.

When the Pools started in 1923/1924 they had no facilities. They contracted with some line elevator companies to handle their grain. UGG and Alberta Pacific handled much of the pool grain at this point. In those fledgling years, UGG provided much assistance to the Pools, financial and otherwise. By the late 1920s, the pool movement was doing well financially, and the Pools started building their own elevators in their respective provinces. Most progressive in this construction was the pool movement in Saskatchewan. As the Pools constructed more facilities, disagreements with UGG over the handling arrangements of pooled grain arose, and the Pools and UGG parted ways from their earlier close working arrangement.

The three Pools banded together to form an international marketing organization to sell their wheat internationally. It was called the Central Selling Agency (CSA). Farmers would consign their wheat to a pool and receive an initial payment. Once the wheat was sold and expenses calculated, the balance remaining was distributed to farmers by means of a final payment. Meanwhile, if a farmer sold his wheat though UGG or a privately owned grain company, he would receive payment in full when the wheat was delivered and the transaction was complete.

Overall, the 1920s were a boom time for the grain industry, primarily in the years after the war and then in the latter part of the decade when wheat prices were high. During this period, Prairie wheat farms experienced unprecedented wealth and profitability. Much marginal farmland was put to the plow so more wheat could be grown. This period also spawned the construction of many grain elevators on miles and miles of newly constructed rail branch lines that crisscrossed the Prairies in anticipation of boundless wheat shipments. The branch lines were built to reach into areas where farmers had a long haul to existing elevators on main lines. New elevators built on the branch lines would collect grain and funnel it to the main line.

This was the apex in the number of grain elevators and miles of rail line in Western Canada. By 1930, there were over five thousand grain elevators across the Prairies. Grain companies with names such as Alberta Pacific, Federal Grain, United Grain Growers, Searle Grain, Home Grain, Gillespie Grain, Reliance Grain, and Bawlf Grain dotted elevator sidings across the West. Canada was a dominant world player, supplying over 35% of wheat exports worldwide.

"Hauling grain to Vulcan, Alberta.", 1928, (CU166071) by Oliver, W.J. Courtesy of Glenbow Library and Archives Collection, Libraries and Cultural Resources Digital Collections, University of Calgary.

This was how farmers hauled their grain to country elevators in the 1920s. The railways had constructed sidings every six to seven miles along their lines. On these sidings the grain companies built their elevators and small prairie towns sprouted. It was felt that this distance between grain elevators was reasonable for a farmer to take in a load of wheat by horse and wagon and return to the farm on the same day. This photo of a line of grain elevators along the rail track was called an "elevator row" and was in Vulcan, Alberta in 1928. Vulcan had one of the longest elevator rows on the Prairies. It was called "nine in a line."

≈

From 1901 to 1931, the population of the prairie provinces increased fivefold due to the influx of new settlers and growth of the wheat economy. In those years, Saskatchewan farmers experienced a boost in wealth the agriculture sector never achieved again. All of this would prove unsustainable over time.

The structure of worldwide wheat exports had changed substantially during World War I. Canada had been a large exporter prior to the war. During the war, wheat production in continental Europe was ravaged and British production was lower because of labour shortages caused by men going off to war. Also, with the Bolshevik Revolution in Russia, wheat exports from the Black Sea area totally stopped and did not commence again until eighty years later when the Soviet Union was dissolved.

After WW I, European and British wheat supplies were replenished, as were stocks in Canada and the United States. Two smaller exporters, Australia and Argentina, were slowly increasing their respective shares in the world wheat trade. Stocks of wheat were quite comfortable in the world by 1930.

Initially, in the mid-to-late 1920s, the Pools through the CSA were quite successful. Saskatchewan Wheat Pool was able to voluntarily contract 50% of the wheat grown in that province. The CSA was internationally marketing 50% of wheat exports from Canada. This would all change with the market crash in late 1929.

GROWTH IN THE RAIL SECTOR AND
DEVELOPMENT OF WEST COAST PORTS

The West grew rapidly in the first two decades of the twentieth century. Advances were made in the railway sector as well. Shippers were frustrated with the monopoly the CPR had on rail movement. The CPR transcontinental line, completed in 1885, ran across the southern Prairies through Winnipeg, Regina, and Calgary; through Rogers Pass; and on to Vancouver. A multitude of branch lines fed into this main line from both the south and the north.

With rail traffic growing rapidly between 1901 and 1905, two new railway upstarts, the Grand Trunk Pacific (GTP) and the Canadian Northern (CN), started building competing transcontinental lines to serve the northern prairies. In 1905 the Canadian Northern line passed through an area about 50 miles east of Edmonton that eventually became the town of Mundare. The rail line and new town were established a little over a mile south of the location of my grandfather's 1899 homestead site.

Both the Canadian Northern and the Grand Trunk Pacific were privately funded. The idea was to build a new northern line from Winnipeg through Saskatoon and Edmonton, through the Yellowhead Pass, and on to Vancouver. These two railroads were contending to complete their respective lines ahead of the other before they reached the Yellowhead Pass. There was room, both financially and geographically, for only one line through the mountains.

Both entities ran out of capital near Edmonton, before either of their rail lines were completed. These lines were geographically close together as they essentially paralleled each other across the northern prairies. In some cases they were less than twenty miles apart. This proximity of the two new lines resulted in them providing competition to each other rather than competition to the CPR, which was some two hundred miles to the south. Neither the GTP nor the CN could individually generate enough traffic to be sustainable in the long run.

Both the GTP and the CN became insolvent during World War I. The federal government had guaranteed their bonds, so it took over their assets and transferred them into a Crown corporation called the Canadian National Railway (CNR). The government wished to see this second transcontinental line completed to the West Coast to provide competition to the CPR.

Prior to the 1920s, most prairie wheat exports had been going east by rail from the Prairies to Thunder Bay at the northwest end of Lake Superior. Here the wheat was loaded onto vessels that crossed Lake Superior and went through Lake Huron to Georgian Bay on the east side of Lake Huron. Then the wheat was unloaded into transshipment terminals that transferred the wheat onto railcars for shipment to domestic markets in Ontario and Quebec. Feed grain was shipped to Eastern Canada in a similar manner.

Wheat to be exported was railed to grain terminals on the St. Lawrence River at locations such as Montreal and Quebec City, where ocean-going vessels would be loaded with export wheat. This was a long, complex, and costly journey. When the St. Lawrence Seaway was completed in 1959, vessels called "lakers" would ship grain all the way from Thunder Bay to the St. Lawrence through the Seaway. The lakers were constructed to exactly fit the size of the locks on the seaway. Though this simplified the long and complex shipment of Prairie grain eastward, this route was still hampered by the fact that shipping was closed on the waterways during the winter months.

Exports through the West Coast ports began growing in the 1920s and grain companies started building export terminals in Vancouver. A second rail line to Vancouver was necessary to provide competition to the CPR in this evolving grain route from the prairies. This more northerly route through the Yellowhead Pass was simpler and much shorter, but still had to deal with the difficult issue of crossing the Rocky Mountains.

As part of the terms of taking over the bankrupt rail assets to form the CNR, in 1925 the government extended the Crowsnest Pass rates to include shipments to the West Coast and made these rates statutory at 1897 levels. This seemed like a reasonable arrangement at the time and resulted in more traffic flowing west to the Pacific. However, the result of fixed statutory rail rates to both Thunder Bay and the West Coast over the next fifty years discouraged the railways from hauling grain since they were losing money with every bushel transported.

The opening of the Panama Canal in 1914 also changed the dynamics of wheat exports from Western Canada. It became feasible for ocean-going vessels to load on the western coast of North America, thereby saving 8,000 miles and the stormy waters around Cape Horn. This resulted in a reasonable and economical route to Britain and Europe. Ocean transport of bulk commodities like wheat is much more economical than overland transport. This also made grain-growing more profitable on the western Prairies, especially Alberta, since it was no longer necessary to make the long land-and-water route eastward to St. Lawrence terminals.

In 1925, the Canadian government built another of their slip form concrete terminals north of Vancouver at Prince Rupert. The government then owned two such facilities on the West Coast and leased them out to grain companies.

Export of grain through Prince Rupert did not evolve as well as it had through Vancouver. In the early 1980s, the Alberta government financed the construction of a large, modern terminal in Prince

Rupert that opened in 1985. It was owned by a consortium of the three Prairie Pools, UGG, Cargill, and Pioneer Grain. Each company's share of the ownership reflected their individual proportion of Prairie grain handled. This new Prince Rupert facility turned out to be a poor and costly decision for the Alberta government. It was never very successful.

The mainstream grain companies were much more anxious to ship grain through their own terminals in Vancouver rather than the joint venture in Prince Rupert. They had been strongly urged and encouraged (some would even say coerced) by the Alberta government to build there as a West Coast shipping alternative to Vancouver. This occurred in the shadow of the late-1970s collapse of the Western grain-handling-and-transportation system, as will be outlined later. The consortium of six companies eventually defaulted on the loan and the Alberta government was forced to take it over. It now remains largely under-utilized, as billions of dollars have been spent improving facilities for grain flow through Vancouver in the past twenty years.

Most Prairie grain is now exported through the West Coast. Unlike grain shipments eastward through the St. Lawrence where shipping only occurs about nine months of the year, the West Coast ports are open year-round. With the growth of export markets in the Orient over the past four decades, the West Coast is now the most economical tidewater destination to service these markets.

THE GREAT DEPRESSION, WORLD WAR II, AND THE WHEAT BOARD BECOMING MANDATORY

I just couldn't believe that you could, in isolation, in the middle of the continent of North America, set up a system where the buyers of the world were going to accept your price, because you've got to meet your customer on common ground somewhere. But the Pools and the CSA visualized this as a method to hold them up and make them pay more money.

Mac Runciman, President, United Grain Growers, 1961–1981

With the onslaught of the Great Depression, the price of wheat fell precipitously. Wheat on the Winnipeg Exchange went from $1.24/bushel in 1929 to $.37/bushel in 1932. The later value would result in the price of wheat being about $.20/bushel at a country point after freight and handling were deducted.

The three Prairie Pools, through the Central Selling Agency (CSA), found themselves owning millions of bushels of wheat worth much less than they had paid farmers for through the initial payment in 1929. Grain merchants usually hedge their grain transactions to protect themselves from price risk. After purchasing a farmer's grain, a merchant will sell a similar amount on the futures market. If the price goes down while the merchant is marketing the grain, this risk will be offset by the "short" position held by the merchant on the futures market. The merchant will be "long" physical stocks of wheat while being short on the futures. These two transactions will offset price risk while a merchant is holding stocks of grain.

When the market crashed, the Pools (through the CSA), found themselves with millions of dollars of losses. They had purchased farmers' grain at pre-Depression prices. As the market fell, they held back on sales, hoping this would drive up the world price. This did not happen since Canada was no longer as large a player in the world wheat market as it had been a decade before. To compound matters, since the Pools were philosophically opposed to the Grain Exchange, the CSA did not hedge its purchases as most merchants would. The value of the inventory of their wheat fell drastically below the price they had paid for it.

Not only was the inventory of wheat the CSA purchased at a high price from farmers in the 1929–1930 pool not hedged, but the CSA opted to actively purchase wheat futures on the Winnipeg Exchange in an unsuccessful attempt to try to prop up world prices. This resulted in them "doubling down" on their losses. They owned physical stocks purchased from farmers in the pool and were buying wheat futures at the same time, resulting in a compounding of their losses as the market continued to fall.

Despite the Pools' dislike of speculators and the Winnipeg Exchange, they were speculating by buying wheat futures and being long not only on their purchases from farmers but also on the Exchange. Wheat prices continued to plummet and the CSA became encumbered with huge liabilities they could not meet.

The three Pools found themselves unable to meet their financial commitments in 1931. Each province financially bailed out their respective Pool with loans. With this government assistance, the Pools became operational again, though they no longer operated pools of wheat purchases as they did previously. They existed as grain companies in their respective provinces, operating in a similar manner as, and in competition with private companies, paying the entire price up front when they purchased wheat from farmers. The federal government took over the entire stock of wheat and the CSA was dissolved. The Pools then resumed their campaign for Ottawa to establish a compulsory wheat board as existed during World War I. They did not want the private grain trade to dominate wheat marketing on the Prairies.

The 1930–1931 pool account was not paid out and remained open until 1936 when the new CWB was formed. In 1936, the federal government paid equalization funds into the 1930–1931 pool account to ensure all farmers received the same price in that year. When the market was falling in 1930–1931, and the CSA was desperately trying to stay alive financially, over the ongoing year the initial price was dropped several times to follow the falling market. This resulted in farmers who had delivered into the pool earlier in the year receiving more than those delivering later in the year. The final equalization payment that came with the closing of the pool in 1936 ensured all farmers who delivered wheat in the 1930–1931 pool year received the same price. The wheat surplus taken over from the CSA was finally dealt with.

To compound the financial woes of the Prairie wheat farmer due to the Depression, the drought years of the Dirty Thirties compounded the hardship. In eight of the ten years between 1929 and 1938, the yield of Prairie wheat was markedly below the long-term average. While the value of agricultural production on the Prairies was near $1 billion in 1926–1927, in 1933 it was $163 million. Consecutive years of below-average rainfall and above-average temperatures resulted in several poor crops. In drier parts of the Southern Prairies there was little or no snow, which, along with low summer precipitation, resulted in a sparse foliar growth. Hot, dry, windy conditions resulted in significant soil drifting. Many farms were simply abandoned. Hardship, both financial and otherwise on Prairie farms, was unprecedented. The price of wheat did not recover until World War II in the early 1940s.

In 1932 the price of barley was so low that by the time the costs of freight, elevation, and handling were deducted there was hardly any money left for the farmer. With dark humour, an unlikely tale was told of a farmer who delivered a wagonload of barley to the local elevator. By the time all the deductions were taken, the farmer found he owed the elevator company money. The manager demanded payment and the farmer said he had no money, but next time he came to town, he was to bring a chicken in payment. The elevator manager agreed. The next week the farmer walked into the elevator office with two chickens. The manager said, "The deal was you owe me one chicken. Why have you brought two?" The farmer replied, "I brought in another load of barley."

As the Conservative government of Prime Minister R.B. Bennett took responsibility for the millions of bushels of wheat taken over from the Central Selling Agency in 1931, the price of wheat continued to diminish through the Depression. World stocks were more than adequate, as importing countries had no funds to purchase wheat due to the Depression. Even though the crops in the 1930s were poor, there were still additional crops of wheat grown annually. This wheat was marketed by the grain companies in competition with the government stocks.

Bennett was familiar with the grain industry, as he was from Alberta and had been a director of Alberta Pacific Grain, a private grain company based in Calgary. He tasked John McFarland, the former general manager of Alberta Pacific Grain, to manage the large government stocks taken over from the CSA and market them to provide a measure of stability to the Prairie wheat market.

It was a difficult assignment. Every time McFarland tried to market some of these stocks, the price would go down and the attempts would be criticized by the trade. Meanwhile, farm organizations led by the Prairie Pools were putting increased pressure on the government to reinstate a wheat board, since many farmers favoured a price-pooling marketing system. Wheat prices remained abysmally low. Unable to dispose of government stocks and criticized by the entire grain trade, both left and right, in 1935 Bennett relented and created a voluntary wheat board governed by three commissioners, with John McFarland as the first chief commissioner. The newly formed Canadian Wheat Board only applied to wheat in the West; was voluntary; and guaranteed initial payments to farmers from the Government of Canada.

Of lasting significance was Bennett's declaration that grain elevators and railways are "works for the general advantage of Canada." This meant that the government or its agencies could control the use of these facilities.

With much controversy, fury, and criticism from opponents, the new CWB Act was passed in August of 1935, just two months before the Conservative government of R.B Bennett was ousted by the Mackenzie King Liberals. Shortly after the election, the new Liberal government called for the resignation of the three commissioners appointed by Bennett and installed three of their own appointees to oversee the new CWB.

King's Liberals did not wish to be in the grain business. One of the staunchest opponents to the government being involved in wheat marketing was Tom Crerar, former president of United Grain Growers and now a senior member of the King Liberal cabinet. Attempts were made to disband the new CWB, but pressure from many western MPs and the Prairie Pools resulted in its continuance. Grain policy drifted from 1935 until the war set in. As the small crops of the drought years continued, the CWB slowly managed to dispose of the stocks inherited from the CSA earlier in the decade.

The year 1937 is notable, as that is when the Prairies experienced their worst wheat crop ever. The average yield of wheat in Saskatchewan was 2.5 bushels/acre that year. Many areas of the province had no seed grain for 1938. The CWB was given the task of gathering seed wheat from areas in the Prairies that had a surplus and distributing it to areas where the drought was so severe that farmers did not harvest enough to provide seed for 1938.

As the world entered the early war years, grain trade was disrupted due to shipping risks from German patrols on the Atlantic. Stocks started to build up again in Canada, and the British once again were becoming concerned about wheat supply. In 1940, the CWB Act was amended and the CWB was given more power, though it was voluntary. Perhaps the most significant amendment was "the board was given power to regulate deliveries by producers at country, mill terminal elevators, and loading platforms".[6] These would be all facilities licensed by the Canadian Grain Commission.

The first CWB delivery permit books were issued in 1940 to ration and equalize farmers' grain sales to both the private trade and the voluntary CWB. This would lead to a structural shift in how wheat was delivered on the Prairies for the next sixty years. Regulation rather than price would determine the flow of grain into the market. In 1942, the CWB was given power over rail boxcar allocation—another action that was to become controversial in future years. These additional powers given to the CWB were a key reversal in the Liberal government's approach to wheat marketing, and were implemented largely to manage wheat supply as the war set in.

These types of infringements on individuals' rights were tolerated by the public under wartime conditions. The problem arose in later years once world economic conditions normalized. These restrictions were not lifted, and the rights of Prairie farmers in grain marketing remained restrictive.

In 1942 during the dark, early days of World War II, the British once again pressed the Canadian government to take control of wheat trade to guarantee Britain a secure supply while preventing hoarding or profiteering. In 1943, to oversee the supply of Canadian wheat and prevent international prices from going up, the Liberal government in Ottawa once again, as in World War I, used the War Measures Act to grant the CWB mandatory power over wheat marketing. Unlike in 1917, this time the CWB mandate only applied on the Prairies and not elsewhere in Canada.

---

6   Morriss, William E. *Chosen Instrument: A History of the Canadian Wheat Board: The McIvor Years*. Reidmore Books. 1987. Page 127.

*It wasn't the arguments of the wheat pool members
that led the government to make the crucial
decision. It was the exigencies of wartime.*

– Mitchell Sharp, *Which Reminds Me: A Memoir.* University of
Toronto Press, 1994. Page 28. [7]

Until this time, the Liberal government had always opposed
a mandatory CWB. It was clearly external events that forced
the creation of the mandatory CWB, not the farm lobby as the
Pools suggested.

As the CWB took over, the futures market at the Winnipeg
Grain Exchange was closed in 1943. Futures markets are a price-
discovery mechanism where grain traders can bid for grain. With the
government taking over wheat pricing, this open-market method of
price discovery was no longer relevant.

There were world food shortages during the war, and even
in Canada foodstuffs were rationed. The government wanted to
encourage food production. British supply of foodstuffs from
continental Europe was severed during the war, and Britain looked
to Canada for supply. Most of the livestock industry in Canada was
in Ontario and Quebec at this time. The government introduced
a feed-freight assistance program to subsidize the shipment of
feed grain from the Prairies to the feed-deficit central and eastern
provinces. This clearly disadvantaged the growth of livestock
production on the Prairies. This program continued long after the
war as the Feed Freight Assistance Program before it was finally
eliminated in the 1980s. This was another policy from Ottawa that
clearly favoured Eastern Canada over Western Canada. Through its
control over grain price, supply, and movement, CWB was key in
ensuring Eastern Canada had sufficient supplies of feed grain at a
reasonable price.

---

7   Sharp was a former employee of James Richardson and Sons and later a
    minister in the Lester Pearson and Pierre Trudeau Liberal governments.
    During World War II, he was a bureaucrat in the federal government prior to
    taking on his ministerial role.

Many of the powers that were given to the CWB by the King government were rooted in controlling food supply during the war. Though they were originally intended to be temporary, these powers remained in effect well after the war ended.

Also interesting to note, it was during this time that rapeseed first came on the scene in Prairie grain production. The war had also disrupted production, consumption, and trade of fats and oils. This was particularly the case for vegetable oils such as copra, where supply from the Pacific area was curtailed due to the Japanese invasion and occupation of the Pacific. At this time, Phyliss Turner was the administrator of fats and oils on the Wartime Prices and Trade Board. As she was searching for alternate vegetable oils she found out about rapeseed and was instrumental in introducing rapeseed production to the Canadian Prairies. Phyliss Turner was also the mother of future prime minister John Turner.[8]

Rapeseed oil was high in erucic acid, which was not a favourable quality for human consumption. Later the erucic acid would be bred out of rapeseed and it would be rebranded as canola, a much healthier vegetable oil.

≈

After the war there was a great need to restore food stocks, primarily in war-ravaged Europe. Many people were close to starvation. To make things worse, Europe experienced a drought in 1946. The British looked to Canada to help rebuild food supplies. Using the CWB as the mechanism for guaranteed price and supply, Canada entered a four-year wheat deal with Britain. It was at a fixed price for the entire four years. As the world focused on rebuilding supplies, prices increased substantially over the next while. Canada adhered to the terms of the four-year deal as part of the war rebuilding effort

---

8   Morris, *Chosen Instrument*, pages 150–51.

in Europe. Western farmers were forced to subsidize the sale by selling wheat at the low price of the original deal. With the price of this sale locked in for the first two of the four years in the late 1940s, it is estimated that Prairie farmers lost $364 million selling wheat below market price to the British while the world price remained much higher. This would be over $6 billion in current funds.

At this time exports of wheat were very important to the Canadian economy. Wheat sales were the largest contributor to annual foreign exchange earnings, with the United Kingdom being the largest purchaser.

It was necessary for the Liberal government in Ottawa to maintain a compulsory wheat board to manage wheat supplies for the discounted price as per the terms of this British agreement. After the emergency powers of the War Measures Act were rescinded, the government allowed the CWB to maintain its authority over system management and throughput by Order in Council under the National Emergency Transition Powers Act.

Restoring food supply in Europe also included restoring livestock products, which in Canada at the time were produced mainly in Quebec and Ontario. The feed-freight-assistance subsidy program continued. Eastern Canadian farmers responded to higher prices by producing more beef, pork, poultry, and dairy products. The government then bought these products from Eastern Canadian farmers at market price and resold them to the British at lower prices to help replenish British food supply. Unlike wheat from Western Canada, when it came to agricultural products from Eastern Canada, *all* Canadians made up the price difference.

> The bottom line was that when it came to financing Canada's contribution to helping Europe rebuild after the war, Eastern Canadian livestock farmers received the new higher market prices for their contribution, while Prairie wheat farmers bore the cost of the low-priced, four-year, locked-in wheat sales administered by the CWB.

During this post–World War II period of European reconstruction, the Canadian government was involved in many affairs in Europe, such as military supervision, infrastructure repair, and food supply. The government wanted to ensure a secure supply of feed grain to Central and Eastern Canadian livestock producers. This was achieved by continuing with the Feed Freight Assistance Program and making use of the CWB to gather supplies of feed grain from Western Canada and sell it to livestock producers in other parts of Canada.

A major change in CWB authority occurred at this time. Amendments to the CWB Act in 1947 made the CWB an agent of the government. Rather than being described as a quasi-independent government body, designed to act as an agent for the grain growers' protection, the 1947 amendments made the specific declaration that "the Board is, for all purposes an agent of his Majesty in the right of Canada, and its powers under this Act may be exercised by it only as an agent of his Majesty in the said right."[9] This was an important distinction and change in mandate, as the main focus of the CWB shifted from acting on behalf of farmers to acting as an agent of the government.

Prior to the 1949 election with King retiring as leader of the Liberal party, King was forced to accept the fact that he could not extract the federal government from the business of wheat. In a 1949 bill strengthening the CWB Act, the Canadian government expanded the authority of the CWB and by Order in Council mandated that the marketing of barley and oats would be taken off the open market and would fall under the authority of the CWB. While CWB marketing of wheat was enshrined in legislation, marketing of oats and barley was added by Order in Council. This difference would prove to be an important distinction in future years. The Prairie Pools and the Canadian Federation of Agriculture were heartily supportive of this action. United Grain Growers, through all stages of this progression, voiced its opposition to placing the marketing of barley and oats under the CWB. All these actions led

---

9   Morriss, *Chosen Instrument*, page 189.

to farmers having no opportunity to market their own wheat, oats, and barley.

Other amendments at this time included authorizing the CWB to regulate deliveries to elevators and railways. The CWB was also given control over interprovincial movement of grain as well as ensuring that sufficient Western feed grain was available for Eastern livestock producers. Prior to this, the CWB exerted this power under the War Measures Act and then the War Transitional Measures Act. Clearly the CWB was being positioned to become a permanent fixture in Canadian grain marketing, but the Liberal government would not admit it.

On the international scene, in 1949 an International Wheat Agreement was signed by thirty-seven importing countries and five exporting nations. Britain was the principal importer and Canada the principal exporter. The agreement was to expire in 1953. To be able to carry out all these national and international commitments made by the federal government, the CWB was empowered to administer them. That responsibility led to the existence of the CWB continuing after World War II.

The three Prairie Pools were ecstatic to see all this power vested in the CWB. They had been asking for it since the 1920s and now they had it. Not so pleased was the Winnipeg Grain Exchange, the private grain trade, and many individually minded farmers who did not wish to have their grain marketed by a government agency.

Somehow flax and rye were not considered to be important enough and remained on the open market. Canola was not a significant crop at the time. It bears repeating that the powers of this mandatory CWB, based in Winnipeg, applied *only* to Prairie farmers. Canadian farmers elsewhere were able to continue to market any crop they produced to anyone they chose without any delivery quota restrictions.

At this time, many farmers favoured CWB marketing. Some of this reasoning was real and some was perceived. Wheat Pool support and promotion of CWB "orderly" marketing was incessant. There were widely told accounts about the way the railway/grain company combine took advantage of farmers in the late 1800s before the Manitoba Grain Act. Farmers often erroneously equated the increase in wheat prices during WWI with the formation of the first wheat board, rather than the supply disruptions caused by the war. The subsequent drop in prices after the war was due to a replenishment of world stocks rather than the elimination of the first CWB.

In the 1920s, farmers experienced the rise of the co-operative wheat pool movement and price pooling, and some favoured this method, believing it gave them more marketing power. During the Depression, conditions on Prairie farms were so wretched that many farmers blamed the capitalist system and looked to wheat boards and price pooling as solutions to their dire finances. Overall, there was a feeling of helplessness and lack of hope in the farm community during this period. Many farmers were simply more comfortable with marketing through a central agency rather than having to research markets themselves. It was a different era back then. Unlike today where technology provides farmers with instant information on grain prices on their cell phones, in the 1930s, 1940s, and 1950s farmers knew little or nothing about the international price of grain.

The farming environment at this time was in great contrast to the present. There were no modern communications. Overall, farmers were not that well educated. Many farms were quite small, and farmers did not feel they had the ability to deal with clever grain company managers. Unlike today's large, modern trucks, in the 1930s and 1940s grain was hauled to an elevator in town by horse and wagon. Wheat was shovelled onto the wagon the night before, and the next day there was the long trip to town. There the farmer would be greeted by a shrewd grain company elevator manager who would at times use unscrupulous means to entice him to deliver to his facility. It was not unknown for alcohol to be involved in making the farmer less observant. Erroneous weights, grade discounting, and other forms of chicanery were common. The farmer might watch the scale balance accurately but not see that a lesser weight was being recorded on the scale ticket. Once the wagonload arrived in town, it was unlikely the farmer would haul it all the way back home if he did not agree with the grade or price offered.

Farmers did not trust the grain companies, and this is where the account of the infamous elevator manager's little black book requires explanation. If an elevator manager short-weighed a load of wheat from a farmer, a surreptitious entry would be made in a small book, documenting the difference between the amount of grain, the grade, or the price the farmer was paid as opposed to what had been unloaded. The elevator would now be long this amount. The next time a special friend of the manager delivered grain, this long amount would be added to this sale and the proceeds of this stolen wheat would be split between them. By transferring the stolen grain onto another purchase ticket, the elevator books would be in balance again.

It was for this reason that surprise audits of grain elevators often occurred. A representative of the grain company would come in the morning and shut the elevator down. He would then weigh the grain in the elevator and match it to the bookkeeping records of the elevator manager. If there was a discrepancy, it was likely that the manager was engaging in the activity explained above and had not yet had the opportunity to even out the long position by selling it on a favoured friend's grain ticket.

The below account describes what farmers were up against when delivering grain to scheming elevator managers. It was these types of activities that led to the distrust of grain companies by farmers.

growers near Abernethy, Saskatchewan, W. R. Motherwell once recalled how he and his neighbors faced a 25-mile trip to the nearest railway delivery point.[2] Even in good weather, that was then a tiresome, 50-mile round trip by plodding horse and wagon — at a time when a 50-bushel wagonload was a good-sized haul. One mile of travel on rough roads for every bushel, 5,000 miles to deliver all the grain that a moderately large farm might produce, the equivalent of a trek to Montreal and back.

And what would the farmer find waiting for him at the delivery point? Too often, his welcome would be from a smiling elevator agent who would sympathetically explain that the farmer should have been there Tuesday to catch the high wheat prices, or that the wagonload of top-grade wheat looked like a somewhat lower grade, or that a larger than expected weight would have to be deducted as dockage to allow for unwanted weeds or other grains mixed in with the wheat. And if none of these assessments were plausible, there was also the unfortunate possibility that all the elevator bins reserved for No. 1 wheat happened to be full, so that the farmer could deliver only into No. 3 bins, at No. 3 prices. As if all this were not enough, farmers had to be on guard for such other tricks as improperly-adjusted scales. One young boy who was sent to deliver wheat, John Martin, had the load weighed at a lumber yard first, then found that the elevator scales showed a suspiciously lower weight. Martin objected, and the agent reluctantly reweighed the load. Apparently presuming that a boy's word could not count against him, the agent calmly and visibly removed two rollers that had been distorting his scales.[3]

Elevator agents — the ones who kept their jobs — were not usually known for being more charitable than their colleagues at the same point, so even where the farmer had a choice of elevators he usually found a similar welcome. The collusion could at times be insultingly blatant. A farmer who decided to reject the grade offered by one agent and to go to the neighboring elevator would soon realize the futility of such effort when he saw the first agent casually stroll outside and hold up fingers to show the second agent what grade had been offered.[4] The farmer, of course, theoretically had the option of taking his wagonload home and trying the tedious round trip some other time — a choice about as realistic as storing the wheat over the winter in granaries he could not afford to build, or stalling creditors who demanded payment now and refused to accept promises that the wheat could be sold for higher prices in spring.

It was anything but surprising, then, that some talked of seizing grain elevators or blocking rail lines or even using a few bullets to attract Ottawa's attention.[5] Even if the amount of abuse by

Fairbairn, Gary. *From Prairie Roots: The Remarkable Story of Saskatchewan Wheat Pool.*

≈

The entire issue of whether grain should be pooled and sold by a central agency such as the CWB or marketed independently by farmers through individual grain companies became divisive in the farm community. The three Prairie Pools and their followers strongly supported collective marketing as administered by the CWB. Other farmers wished to maintain their marketing independence. Supporters of the CWB referred to this marketing as orderly marketing, suggesting the CWB as a monopoly central agency marketer, with the use of delivery quotas and grain pools, was more "orderly" than the implied "chaos" of the free market system.

This was mandatory collective marketing of farmers' wheat, oats, and barley as administered by the CWB. In addition to use of the term orderly marketing, proponents of this system also referred to it as single-desk marketing. This meant that all Western grain was marketed through one agency and buyers would have to come to this single desk to purchase grain from Western Canada. Grain merchants with Western Canadian grain would not be competing against each other to sell Canadian wheat in the world market.

Emotions and divisions ran high between farmers on opposing sides of the method of marketing wheat. In the 1920s when the Wheat Pool movement was in its infancy, the strongest believers of pooling and the co-op method of marketing felt all farmers should go on strike and stop delivery of all wheat to independent grain elevator companies.

In Mundare, Alberta, in 1934, an organized protest occurred where the most radical leftist farmers associated with the Communist movement attempted to stop any wheat deliveries to the elevators. A dispute occurred when a wagon box of wheat was tipped over and a police officer and several farmers were assaulted. Some of those most responsible were charged, found guilty, and incarcerated for their offences.[10]

---

10 Appendix 1 - Potrebenko, Helen. *No Streets of Gold, A Social History of Ukrainians in Alberta*. New Star Books. 1977. Pages 221–223.

THE CWB AND PRAIRIE WHEAT POOLS AT THEIR ZENITH

In the postwar period, the federal government vested complete power and control over the Western grain-marketing-and-transportation system into the hands of the CWB. For wheat, oats, and barley, the CWB made all export sales, all sales to domestic processors, and all domestic feed-grain sales. They controlled the interprovincial movement of these grains and controlled the flow of grain by farmers into country elevators utilizing delivery quotas. They arranged for railcars and decided which country elevators would receive them, how many they would receive, and which grain was to be loaded onto them. They booked laker freight on the Great Lakes and controlled the flow of grain into the various coastal shipping terminals on both the West Coast and the St. Lawrence. They arranged for vessels to ship the grain and determined at which terminals the vessels would be spotted. The entire supply chain from the farmer's grain bin to the vessel taking on grain was controlled by the CWB. Given this level of power, control, and authority, the CWB was now at its zenith.

All the commercial decisions were taken out of the realm of the marketplace and placed into the hands of a government-appointed central-planning agency with no investment in the system. The only thing the CWB could not control was the world price of wheat.

*If you reduce a dog's chain one link at a time every few days until
his chain is so short he won't be able to move, he will never resist
because he is conditioned to the loss of his freedom slowly over time.*

Author unknown

The CWB operated under federal legislation. Ottawa appointed the five commissioners who ran the CWB. They oversaw grain sales and movement from the Prairies without influences from competition or market forces. It is noteworthy that while Ottawa established and oversaw the CWB, it did not pay for it. Farmers paid for the costs of operating the CWB with deductions from grain sales revenue. Though it was completely controlled by Ottawa, farmers paid for it and had no say in how it operated. The CWB was not a subsidy.

Country elevators and port terminals were owned by the grain companies. Railroad locomotives, most grain cars, and trackage were owned by the railroads. As commanded by the CWB, the system operated in a supply-push rather than demand-pull manner. Market forces played no role in this system. In the post-war economy of the late 1940s and early 1950s, wheat production increased worldwide as socioeconomic conditions stabilized and farmers responded favourably to increasing prices. By the mid-to-late 1950s, there was a world surplus of wheat once again.

Other countries, particularly the United States, increased grain production after World War II. Canada no longer dominated the worldwide wheat export market with over 40% of the market as it had in the prewar and war years. Europe was increasing production and no longer needed to purchase as much wheat. The United States, Australia, and Argentina were becoming strong competitors in world wheat trade.

Canadian wheat stocks increased. Farmers had little cash flow and were forced to build more storage facilities on their farms. The government brought in a temporary storage program whereby they paid grain companies to store grain. Grain-storage annexes were built by the line elevator companies to capitalize on this program.

The Pools thrived on this storage program, as they had the largest storage capacity in their respective provinces.

Harvest on the prairies in the 1950s. My father, Mike Motiuk, harvesting wheat swaths with his Case SP12 combine.

To provide some cash to farmers during this slow period of wheat sales, in 1957 the government introduced the Cash Advance Program. This was a short-term loan that farmers could obtain from the federal government that was secured by the wheat farmers were waiting to sell. This bridged their finances between harvest and the time when the CWB opened a delivery quota. When wheat was eventually delivered, the funds advanced earlier would be deducted from the settlement.

The CWB administered the Cash Advance Program on behalf of the federal government. In the early 2000s, when the Harper government was attempting to amend the CWB, administration of the Cash Advance Program was removed from the CWB. The federal government assigned administration of the Cash Advance Program to the Canadian Canola Growers Association and recently the Alberta Wheat and Barley Commission was empowered to handle these advances as well.

As we moved into the 1960s, wheat prices remained stagnant and sales opportunities poor. Stocks of wheat were high. Canada was slowly declining in its position as a major worldwide wheat exporter. A new oilseed crop called rapeseed was gaining attention and acreage and would soon revolutionize prairie field crop production. Though rapeseed was marketed on the open market by grain merchants, in 1962 the CWB took control of its movement, putting deliveries under a quota system and scheduling shipments.

In 1967, legislation was passed to make the CWB permanent. Until then, enabling legislation had been passed every five years, requiring continual renewal. The CWB was now enshrined in federal legislation, alongside their system control and marketing of wheat, oats, and barley.

In 1970, the three Pools and UGG combined to form XCAN Grain Ltd. to market non-CWB grains abroad (canola, flaxseed, and rye). In 1973, UGG left this arrangement and went on its own. XCAN later became the largest marketer of canola from Canada on behalf of the three Pools.

A game-changing event for the western grain industry occurred in 1972 when the three Pools took over Federal Grain. Federal Grain was the largest non-farmer-owned grain company on the Prairies, with 15% of the market. At this time, the Pools had about 50% in each of their respective provinces and UGG was under 15% of the Prairie market. Federal Grain had just taken over Searle Grain and Alberta Pacific Grain in 1967. It had a very old elevator system, with many small elevators at many delivery points often located on little-used branch lines.

This transaction boosted the Pools to 65% of the business in the West. More important than the old small elevators was the 15% of car allocation that the Pools received with the purchase. Now the Pools could fill up all the small, old Federal Grain elevators with grain and collect storage from the CWB on this grain. They could then direct their 65% car allocation to newer and more modern facilities at points where there were many elevators and a strong competitive environment. Meanwhile, at many small points where the Pool was now the only buyer, the elevator would stay full and collect storage, and the farmers there would be told, "Sorry, this elevator is full." The Pool would then shift the blame by saying the railroad was not supplying cars to this point so the farmer would have to haul his grain further to a main line elevator.

It is important to note that federal taxation policy favoured the Pools. While other grain companies including UGG had to pay tax on their income, the Pools directed all their income to patronage dividends in the name of their customers. It was not recognized as income for the Pools by federal tax legislation. At year-end, individual farmers would get a slip from their Pool that indicated what their patronage dividend was for the year and the amount of tax, if any, they individually owed. In the meantime, the Pool kept all the money for operations, calling it equity. Essentially there was no tax paid by the Pool on their income, and this federal tax concession gave them a tremendous financial advantage. Later, in 1981, former Federal Grain vice-president Gus Leitch spoke of the taxation disadvantage faced by private grain companies. "Saskatchewan Wheat Pool could build an elevator with cash that would have gone to Ottawa if it was a private company. That really hammered us."[11]

In private discussions with more recent UGG directors who were familiar with the Federal Grain deal and had heard the story directly from UGG directors present at the time, Federal Grain wanted to sell to UGG and kept encouraging UGG to buy it. This would have doubled the size of UGG. However, the grain business was a risky investment in 1971. Canada's exports of wheat were down

---

11   Fairbairn, *From Prairie Roots*, page 203.

with no clear sight of improvement. Grain handlings were down and wheat production was being discouraged. In this environment, UGG decided it was too risky, with differing opinions voiced by the then-seated directors. UGG ended up passing on the deal.

The three Pools together were so much larger that the risk was much less for them, so they went ahead with the purchase. In future discussions around the UGG board table, members lamented the decision UGG had made and the missed opportunity. Shortly after the Pools bought Federal Grain came the boom years of the 1970s. The Pools eventually did well with their purchase.

By the early 1970s, the Pools, led by the SWP, were at the peak of their power and influence over the Western grain industry. They were quite profitable, largely because they did not pay income tax and claimed their patronage dividends as equity on their balance sheets. They now had 65% of the grain handle. They were strongly supportive of the CWB, which was in control of the hyper-regulated grain industry. They had a powerful socialist-minded president by the name of Ted Turner. The similarly minded ex-president, Charles Gibbings, was now a commissioner of the CWB.

The Pools positioned themselves as spokesmen for Prairie farmers. They claimed a membership that at times surpassed the number of farmers according to census data. Anyone who bought a $5.00 membership and delivered a load of grain to a Pool once in a lifetime was counted as a member, and the lists were not always updated after deaths and retirements. The Pools purported to speak for all of them, alive or not. As most farm groups had to sell memberships annually to finance their lobby, the Pools simply took finances from their lucrative grain handle and used them to further their policy position, which revolved around CWB marketing and maintenance of the Crow Rate.

The Pools, primarily SWP, considered themselves "a cut above the rest" when it came to farm policy positions. In 1974, when speaking before a Canada Grains Council meeting in Edmonton, Minister Otto Lang mused that perhaps it was time to look at the Crow Rate and compensating the railways for hauling grain. SWP President

Ted Turner took umbrage with these comments and "felt that Lang had adopted an unnecessarily public and belligerent posture with his speech before the Grains Council instead of first pursuing the idea in private with the Pools."[12]

In a meeting of the Western members of the Canadian Federation of Agriculture, SWP Vice-President Charlie Gibbings stated: "That is the way Sask Pool wants it, and if it's not going to be that way, we'll play our Quebec card."[13] And that was exactly what they did in 1983 to convince Ottawa to pay the Crow Benefit to the railroads, not producers, in spite of Gilson's recommendations. (For an explanation of Gilson's recommendations, see chapter entitled 'The Battle over the Crow' page 115.)

Though the Pools touted themselves as being democratic, there were restrictions on who could become a farmer delegate in their governance structure. In 1968, the SWP passed a bylaw "requiring all members to have shipped all their grain and livestock through Pool facilities for three years before being eligible to be delegates."[14]

≈

Deficiencies were starting to show in the transportation system as the railways were losing money hauling grain under the Crow Rate. They were not providing sufficient boxcars to move grain, and there were complaints that available cars were not being distributed fairly. The boxcar fleet was declining and not being replaced. "The number in use fell by half in the 10 years ending 1973."[15]

---

12 Fairbairn, *From Prairie Roots*, pages 220–221.

13 Earl, *Mac Runciman*, page 133.

14 Fairbairn, *Prairie Roots*, page 173.

15 Baron, Don. *Canada's Great Grain Robbery*. Don Baron Communications. 1998. Page 106.

"Terwilliger and Wolfe grain elevator and construction crew, Countess, Alberta". [ca. 1914-1920], (CU1217661) by Unknown. Courtesy of Glenbow Library and Archives Collection, Libraries and Cultural Resources Digital Collections, University of Calgary.

The above photo depicts a boxcar spotted at an elevator. There were large doors on either side of the car. These doors had to be boarded up (coopered) before they could be loaded. Loading these cars was a tedious process, as there were no hatches on the top through which the grain could fall. The grain was dropped in through the side door by spouts extending from the grain elevator. The force of the grain coming from the top of the elevator through the spout would cause the grain to shoot out to the corners of the boxcar, thus loading the boxcar in a very uneven manner.

Prior to loading the boxcar, the elevator manager would have to clean out the refuse left in the car after it was last unloaded at port. This would include the used wooden planking or reinforced cardboard that was used to cooper the car doors from the last load. As a teen, I worked coopering cars for Bill Fedoruk, a Searle grain elevator manager in Mundare in the late 1960s.

These boxcars were also difficult to unload at port terminals as there were no openings in the floors. In earlier days, the grain was shovelled by hand out of these cars. Later, a boxcar-unloading device was devised by C.D. Howe's engineering firm that was installed in the terminals. It would take a loaded and uncoupled boxcar, pick it up, and lift and tilt it from side to side to get the grain out. It was called the Cadillac Dumper. It would have been much easier to ship grain in larger and more efficient hopper cars that were precisely loaded through hatches on the top and allow gravity to unload them through slides at the bottom of the hoppers on the car.

The railways were unwilling to make the investment in hopper cars if they were losing money hauling grain under the Crow Rate. Use of awkward and inefficient boxcars was the main conveyance for Western Canadian grain into the 1970s, over 50 years after the above photo. The technology of hopper cars had long been available, but not for prairie grain!

Bill Fedoruk in his office at Searle Grain in Mundare in the early 1960s. This is the view I would see as I sat with my father in chairs across the desk from Bill as he made out cheques for wheat my father had delivered.
Photo courtesy George Skulsky

≈

The CWB system of marketing did not provide market signals to meter the flow of grain into the system, so the CWB devised the delivery-quota system to equalize wheat-delivery opportunities among all farmers. This was a regulator's dream and a farmer's nightmare. The quota system as designed by the CWB was based on a farmer's total acreage. It applied to all off-farm deliveries of grain to any buyer or destination, only excluding farm-to-farm sales. It even applied to canola, flax, and rye, which were marketed by private grain companies and not the CWB. The CWB claimed it needed to have control over all grain movement so the handling system would not get congested with other grains and oilseeds and thus hinder the movement of wheat.

One must know how the system worked to fully understand how it affected farming both agronomically and regionally on the prairies. Each year, all farmers had to apply for their CWB delivery-permit book (as it was called) at the beginning of the season. A crop year ran from August 1 to July 31 the next calendar year. The permit book was a staple on every prairie farm. In it you had to stipulate the grains grown on your farm and the number of acres of each; the acreage of land being summer fallowed; and the acreage of non-arable land. Then for the grain you wanted to sell, you had to assign "quota acres."

When you applied for your annual permit book, you also had to declare what was to be your "delivery point." This was usually the elevator siding in a town or village closest to your farm. When delivery quotas were opened, you were only able to deliver grain to elevators at that point. If elevators were full at that point, you had to wait for railcars to come and create space. You could apply to change your delivery point, but if you did so then you could not deliver to the original point. So why would anyone haul grain to a further location at a higher cost? Also, it was widely discouraged to change delivery points, and an application to change your delivery point had to be made to the CWB, which was not expeditious in dealing with these requests.

Since the Pools were so large after the Federal Grain purchase, they had many delivery points where they were the only company. They did well with this type of permit-book system as designed and administered by the CWB. The CWB said they needed this type of rigid control over grain deliveries to properly manage the supply chain of grain movement.

To understand how a permit book worked, let us look at a farm with three hundred arable acres in the moist, parkland belt of the prairies where our farm is located. In the 1960s, we would have followed a three-year rotation, with one hundred acres of wheat seeded in year one, one hundred acres of oats in year two, and one hundred acres of land summer fallowed in year three.[16] If we used the oats for livestock feed, and there was no production on the summer fallow, there would be three hundred acres available for quota allocation to wheat.

With wheat yields of about 45 bushels/acre, 100 acres of wheat would produce 4,500 bushels of wheat. With 300 total arable acres available for quota assignment it would take a 15 bushel quota to sell all the wheat from the 100 acres seeded. One can see that fallow acres and feed outlets for other crops were necessary to free sufficient quota acres to get the necessary amount to sell all the wheat, the main cash crop.

In the drier regions of the southern Prairies fallowing was used as a means of moisture conservation. In these arid regions, two years of precipitation would often be required to grow a wheat crop. A

---

16 Summer fallow was a practice whereby the land was not sown for a year and left "fallow." The myth was that you had to leave the land to "rest" so it would grow a better crop next year. In fact, during the fallow year inert nitrogen sources within the soil structure would convert to nitrogen useable by plants and thus result in a better crop the following year. This worked for a while after the land was homesteaded and freshly broken. As the land was sown over more years these inert nitrogen sources became depleted and chemical fertilizers would have to be added to grow a healthy crop. In the southern Prairies where rainfall is limited, this practice is still used to conserve two years of moisture to get one crop.

typical 300 acre farm would follow a 2-year rotation with 150 acres of wheat and 150 acres of fallow annually. With a typical yield of 25 bushels/acre on these drylands, a wheat crop of 3,750 bushels would be grown annually. With 300 assignable quota acres, a quota of 12.5 bushel/acre would empty the bins for the year.

The higher production area in the more northern and moist regions of the prairies were penalized by this quota system and more intense production was stymied since more quota was needed to sell the entire crop. Quotas were usually opened in three bushel increments, and an additional three bushel quota would not be opened until farmers across the prairies delivered their first three bushel quota. In years when the CWB successfully sold the entire crop, at the end of the crop year there could be an "open quota" and farmers could deliver any grain they had left.

The CWB permit book was the prairie farmers' authorization to sell grain. On farms that had livestock, some of the grain production was used as feed so less quota would be needed to sell all the commercial grain produced. On the other hand, productive grain farmers in the more fertile northern parts of the prairies often had difficulty selling all their grain as they would not have enough quota. CWB quotas did not distinguish if you were in a low producing, dryland part of the southern prairies versus a moist, fertile, highly productive area in the northernly parkland belt.

Elevator managers knew which farmers would have surplus quota, and which farmers needed more quota to sell their crop. Towards the end of the year, industrious managers would match these farmers up, and the grain farmer would sell grain utilizing the livestock farmer's permit book. CWB regulations did not allow this flexibility. However, an astute manager could assist a good grain farmer by finding him unused quota. This would then result in increased grain handle for that facility. In the 1960s I personally witnessed this type of activity when visiting the grain elevator office with my father.

Leaving land fallow so the acres could be used for quota assignment had a detrimental effect on the soil from an agronomic perspective. Dr. Don Rennie at the University of Saskatchewan conducted research on leaving land fallow and concluded that it is harmful when exposed soil is left to the elements and subject to wind and water erosion. "His work challenged long-accepted agricultural practices such as summer fallowing, the process of leaving land plowed but unsown to allow nutrient levels to recover. Rennie argued the practice was destroying crop land and recommended that farmers leave stubble on the ground to control erosion."[17]

This conclusion is supported by Les Henry, also an acclaimed soil scientist from the University of Saskatchewan. Over the years, significant soil deterioration had occurred. The CWB quota system encouraged the practice of summer fallow acres to provide more opportunity for wheat delivery.

"Dust mulch, Lethbridge, Alberta.", [ca. 1930s], (CU1126939) by Unknown. Courtesy of Glenbow Library and Archives Collection, Libraries and Cultural Resources Digital Collections, University of Calgary.

17   CBC News. *Saskatchewan soil erosion pioneer dies.* December 24, 2007,

Excess cultivation as in summer fallow results in soil degradation. Surface crop residue is destroyed and organic matter is lost. The soil is then left vulnerable to wind and water erosion as the previous photo illustrates.

In the attempt to provide equal delivery opportunities for all farmers, the metrics of this quota-delivery design penalized those who produced more on their land. Figuring out how to navigate your way through the CWB delivery-quota system was necessary to sell the grain grown on your farm.

When a farmer delivered grain to anywhere other than another farm, he was required to provide his permit book so the purchaser could enter the delivery against the available quota. The CWB would send inspectors to country elevators to make sure every farmer's permit book was up to date and check that each farmer had not delivered more than his quota allotment. There were penalties for both farmers and grain buyers if they violated the allowable quota delivery.

The following photos of a 1970/1971 CWB delivery permit book are provided courtesy of Saskatchewan farmer Hubert Esquirol. The oppressive nature of the permit book that only Prairie farmers were bound by is best explained by the following:

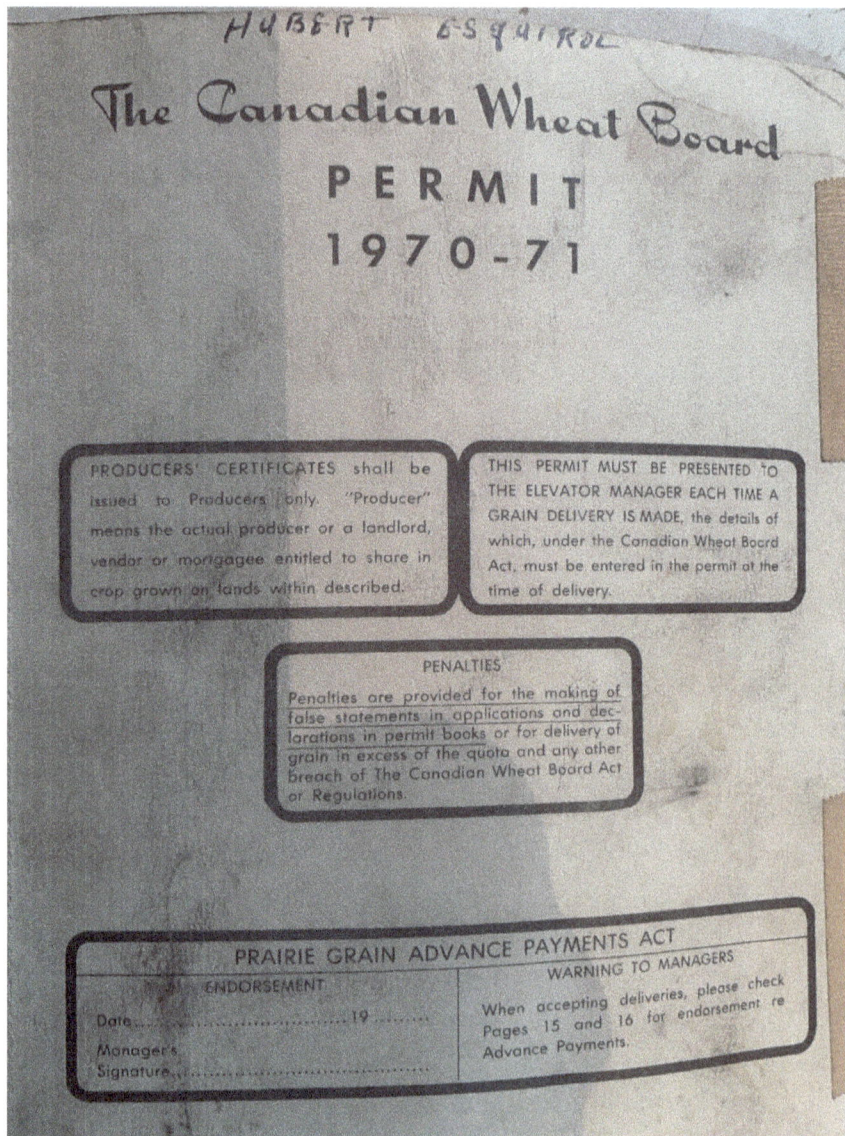

The Canadian Wheat Board
PERMIT
1970-71

HUBERT ESQUIROL

PRODUCERS' CERTIFICATES shall be issued to Producers only. "Producer" means the actual producer or a landlord, vendor or mortgagee entitled to share in crop grown on lands within described.

THIS PERMIT MUST BE PRESENTED TO THE ELEVATOR MANAGER EACH TIME A GRAIN DELIVERY IS MADE, the details of which, under the Canadian Wheat Board Act, must be entered in the permit at the time of delivery.

PENALTIES

Penalties are provided for the making of false statements in applications and declarations in permit books or for delivery of grain in excess of the quota and any other breach of The Canadian Wheat Board Act or Regulations.

PRAIRIE GRAIN ADVANCE PAYMENTS ACT

ENDORSEMENT

Date.........................19.........

Manager's
Signature.......................................

WARNING TO MANAGERS

When accepting deliveries, please check Pages 15 and 16 for endorsement re Advance Payments.

This was the cover of the permit book. Note the instructions as to how it should be used and the penalties for noncompliance.

The Canadian Wheat Board

PERMIT BOOK APPLICATION – 1970-71 CROP YEAR

00000A

ACTUAL

SIGNATURE OF PRODUCER ACTUALLY ENGAGED IN PRODUCTION OF GRAIN	DATE COMPLETED
	June 29 1970

IS APPLICATION FOR WHEAT INVENTORY REDUCTION PAYMENT BEING MADE? YES [X] NO [ ]

ALTERNATE DELIVERY POINT CAVALIER

FOR OFFICE USE ONLY 7133409

LANDLORD, VENDOR, MORTGAGEE ENTITLED TO SHARE IN THE GRAIN GROWN ON THE LANDS DESCRIBED.

	LAND UNCHANGED FROM 1969-70 PERMIT						
PART	SEC.	TP.	R.	M.	ACRES	SFX
B	SE¼	22	49	18	3	160	A
C	BROWHELL LC	N½	22	49	18	3	320	B
Edam	Sask.						
D	WEBER HARU	N½	3	48	18	3	320	C
	SE¼	4	48	18	3	160	C
Edam	Sask.	NW	32	47	18	3	80	C
E							
	TOTAL	1040					

USE THIS AREA TO LIST LAND CHANGES FOR THE 1970-71 CROP YEAR

ADDITIONS

1969-70 FORMER PERMIT HOLDER

ADDITIONAL LAND OPERATED IN THE 1970-71 CROP YEAR

PRODUCER'S IDENTIFICATION NO.

PRODUCER'S IDENTIFICATION NO.

TOTAL 1970-71 FARM ACRES

SUBTRACTIONS

PRESENT 1970-71 PERMIT HOLDER

1969-70 LAND NOT OPERATED IN THE 1970-71 CROP YEAR

PRODUCER'S IDENTIFICATION NO.

PRODUCER'S IDENTIFICATION NO.

FORM NO. DF-1

TOTAL 1969-70 LAND NOT OPERATED IN THE 1970-71 CROP YEAR

PRODUCER OF GRAIN GROWN ON THE LANDS DESCRIBED

0 2357284 5   29060   00000

PERMIT NUMBER

GUIROL HUBERT
23572845

P.O. BOX NO   PROV.   DELIVERY POINT

467229412

4 M   118 SASK   EDAM   1

CALCULATION OF 1970-71 QUOTA ACRES

1970-71 SUMMERFALLOW — **540**

ADD 1970-71 SEEDED ACRES OF OTHER ELIGIBLE CROPS — + —

ADD 25% OF 1969-70 NEW BREAKING — + —

ADD 25% OF 1969-70 SUMMERFALLOW — **55** ▷ **595**

ADD ACREAGE INCREASE IN 1970-71 PERENNIAL FORAGE — + —

OR
LESS ACREAGE DECREASE IN 1970-71 PERENNIAL FORAGE — — —

	WHEAT	S.W.S.	OATS	BARLEY	RYE	FLAX	RAPE	
ASSIGNED ACRES	495						100	H 595
1970-71 SEEDED ACRES				140			80	TOTAL ASSIGNABLE ACRES
1970-71 TOTAL QUOTA ACRES	A 495	B	C	D 140	E	F	G 180	

SEEDED ACRES

	1969-70	1970-71	INCREASE	DECREASE
WHEAT (EXCEPT DURUM & S.W.S.)	460	I —		460
DURUM		J		
SOFT WHITE SPRING		K		
OATS	60	L	60	
BARLEY	6	M 140		
RYE	14	N		
SUMMER-FALLOW	220	O 540	320	
NEW BREAKING	—	P —		
PERENNIAL FORAGE	—	Q —		
ADJUSTED SPECIFIED ACRES	760			
FLAXSEED	—	R —		
RAPESEED	—	S 80	80	
MISC. CROPS	—	T —		
UNCULTIVATED LAND & PASTURE	280	U 280		
TOTAL FARM ACRES	1040	V 1040		

SEEDED ACRES FROM 1969-70 PERMIT BOOK

WHEAT	OATS BARLEY RYE	SUMMER-FALLOW	NEW BREAKING	PER-ENNIAL FORAGE	FLAX RAPE	MISC. CROPS	UNCULT. LAND & PASTURE

The previous two photos are the inside of the permit book. All the land locations, acreage, and land ownership had to be declared. This was so that multiple farmers would not try to use the same land twice in acreage assignment to obtain extra quota. On the top right of the photo on page 69 is a list of the land farmed along with the owners, and their addresses.

On the bottom right-hand side of the photo on page 70 the amount of land sown to each crop or summer fallowed had to be declared (seeded acres). On the top right-hand side of the photo on page 70 was the acreage assigned to the delivery of each crop (quota acres). Note the sum acres of the land farmed on the top left-hand side (1040) of the page 69 photo matches the total farm acres (1040) on the bottom right-hand side of the page 70 photo.

In the page 70 photo you can also see that summer fallow acres on the Esquirol farm went from 220 acres in 1969/1970 to 540 acres in 1970/1971. Acres seeded to wheat went from 460 in 1969/1970 to zero in 1970/1971. This was a direct consequence of the inability to sell wheat off the farm in those years due to lack of sales made by the CWB. Inaction by the CWB negatively affected the flow of both grain and cash for each farmer.

DATE	COMPANY NAME	DELIVERY POINT AT WHICH GRAIN ACCEPTED	QUOTA	GROSS BUSHELS DELIVERED FOR STORAGE OR SALE	NET BUSHELS ACCEPTED ON QUOTA	CUMULATIVE TOTAL DELIVERED	MANAGER'S INITIALS
Aug 25	Pool	CAVALIER	2	495	185		J.J.
" 25	"	"	2		173	358	J.J.
" 25	"	"	2	261	172	530	J.J.
" 26	"	"	2		171	707	J.J.
" 26	"	"	2		159	866	J.J.
" 26	"	"	2		120	986	J.J.
Jan 11/71	Nat	Edam	3		345	1331	J.J.
Jan 28	"	"	4		595	1926	J.J.
Mar 18/	"	"	5		65	1991	J.J.
" 19	"	"	"		44	2035	J.J.
" "	"	"	"		24	2059	J.J.
June 3/71	"	"	6		240	2299	J.J.
"	"	"	"		461	2760	J.J.
" 17/71	"	"	7		334	3094	J.J.
"	"	"	"		167	3261	J.J.
" "	"	"	"		226	3487	J.J.
July 26/71	"	"	8		98	3585	J.J.
" "	"	"	"		362	3947	J.J.
"	"	"	"				

This is the page where all details of grain entries over the year was made. It was to ensure that nobody delivered more than their quota. It also ensured that deliveries were only made to the designated delivery points.

THE CANADIAN WHEAT BOARD
ALTERNATE DELIVERY POINT AUTHORITY

Producer _____ Hubert Esguirol _____ IDENTIFICATION NO. | 0 | 0 | 2 | 3 | 5 | 7 | 2 | 8 | 4 | 5 |

The specified number of acres on which deliveries of grain may be made in the 1968-69 crop year are

_____ 220 _____ acres at Delivery Point _____ Edam _____ and _____ 540 _____ acres at Delivery Point

_____ Cavalier _____

Deliveries of grain made at each Delivery Point must be recorded in the Permit Book, along with a notation, designating the Delivery Point at which individual delivery is made. Deliveries of grain may not be made at any other Delivery Point without proper authority from The Canadian Wheat Board.

This shall not be considered as permission to over-deliver the established quota. Deliveries at each Delivery Point must be within the quota at each Delivery Point based on its respective specified acreage, but where minimum quotas are set by the Board based on seeded acreage, such minimums are the total deliveries under the Permit Book, and not total deliveries at each of the above Points.

This endorsement shall be attached to the front inside cover of the permit book which shows the above mentioned Actual Producer's Identification No. and shall form a part of said permit book, and cancels all specified acreage figures heretofore fixed in respect of said permit book.

DATE _____ SEP 6 1968 _____ THE CANADIAN WHEAT BOARD per _____

Elevator Managers — Please check recorded deliveries made to each point listed before accepting new deliveries

This is from Hubert's 1968/1969 permit book where he had to request permission to be able to deliver grain at two delivery points, Edam and Cavalier. His delivery acres were then split to become 220 acres at Edam and 540 acres at Cavalier. Deliveries outside these parameters were not allowed. A similar request had to be made if you wanted to change your delivery point assignment.

Just like a Canadian citizen needs a passport to travel to a foreign country, Western farmers required a CWB permit book to sell their grain. Farmers in other parts of Canada were free to sell their grain directly to any grain merchant without a quota book.

Once we began continuous cropping in 1977, we no longer had summer fallow acres in our permit book. Additionally, we became more productive per acre with increased fertilizer and herbicide usage. The CWB acreage-based quota system created all sorts of challenges for us in terms of finding ways to sell our crops.

> The CWB quota system also applied to grain deliveries beyond the line elevator system. Quotas were applied to canola, flax, and rye deliveries, which the CWB didn't even market.

Advances in tillage, seeding technology, crop varieties, and fertilizer and herbicide use had rendered the practice of summer fallow obsolete in the higher rainfall parts of the Prairies, except for the fact it helped you deliver your grain as controlled by the CWB quota system.

The following table depicts the declining acreage of summer fallow on the prairies over the past 40 years.

Summer fallow acres on the prairies in millions of acres

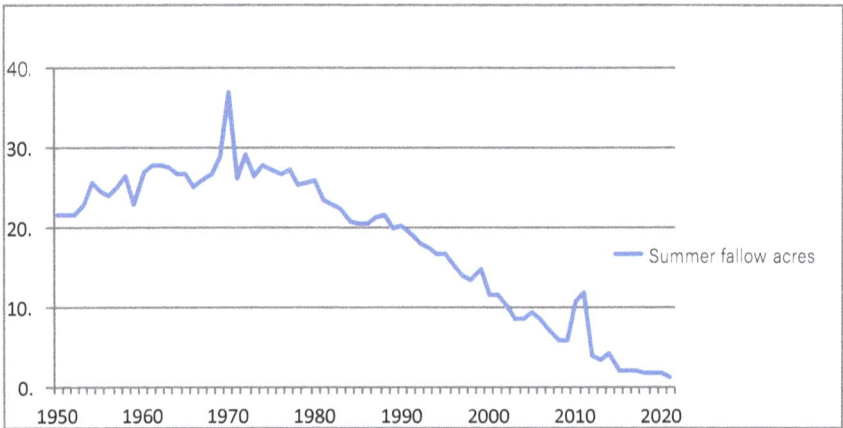

Later, in the 1990s, the CWB introduced a contracting system as an option to the old acreage-based delivery quotas. Summer fallow acres were declining. Over time, significant economic revenue had been lost by farmers (and the Western economy) as the practice of summer fallow continued while more acres of crop could have been grown annually. But summer fallow acres in your CWB permit book were necessary to provide more opportunity to sell the crops grown on your farm.

Quotas were also applied to deliveries of all grains to processing facilities and feed mills. Glenn Keddie, who operates a feed mill in Grande Prairie, Alberta, recounts buying feed grain from farmers and having to fill in the amount of the delivery in their CWB permit book.

Even though no export grain went into these facilities, the CWB maintained that they must ensure equal delivery opportunity for all farmers. Quota, not price, was the mechanism for metering grain deliveries. Years later, when canola became a mainstream crop, the CWB placed quotas on canola deliveries even though the CWB did not market it. A delivery quota was even imposed on deliveries to canola-processing plants that never utilized a country elevator or any export or rail capacity. These authoritative actions demonstrate the encroachment of CWB control throughout the system.

The government had earlier designated all grain delivery and processing facilities to be operated for the greater good of Canada and gave the CWB the power to apply these quotas even when it did not involve export wheat, oats, or barley. At this time, it was even illegal to ship any grain across a provincial border to an elevator, feed mill, processor, or cattle feedlot. Penalties were imposed if these rules were violated.

The CWB paid all farmers the same price, as all the grain sales were pooled. Upon delivery of grain, each farmer received an initial payment, which was about 75% of the market price. The crop year ran from August 1 to July 31 of the next year. For all the grain delivered in this twelve-month period, a farmer received the initial payment. Then from July 31 to December 31 the grain sales would be completed and the accounts for that crop year would be finalized. In January, a farmer received a final payment from all the sales from the pool account, the final amount depending on how profitable the sales were. So, for a crop planted in May 1965, the crop year for deliveries would be August 1, 1965 to July 31, 1966. The final payment for that grain would be received in January 1967, twenty months after the crop was planted. Without knowing what the final payment might be, it was difficult to complete any budgeting or financial planning on a farm.

≈

Another reason the marketing system was becoming dysfunctional was the rail-transportation system for grain. The Crow Rate mandated railways to ship grain at the fixed rates set earlier in the century. Over the years, due to inflation and increasing costs, the railways were losing more and more money on every bushel they hauled. The railroads were not eager to haul grain. Numbers of railcars for shipping grain were always in short supply, and the CWB rationed them out among shippers.

The oversight and control of railcar allocation by the CWB became controversial shortly after the CWB took over this responsibility in the early 1950s. Former Manitoba Premier John Bracken was hired to review the way the CWB was allocating cars and to suggest a fair and transparent system. This became known as the Bracken Formula and was predominantly based on historical handling percentages.

The CWB controlled the movement of grain in, out, and through the system. They calculated their car allocation to individual grain companies based on each company's historical handling percentage. This essentially locked in a company's allotment, or market share, over time. If you had 25% of the Prairie grain delivery market share, you received 25% of the cars. Pricing, efficiency, and customer service were not considered. This made it nearly impossible for an individual elevator company to increase its ability to handle more grain year over year.

As we shall see later, CWB car allocation remained a controversial issue until the CWB was disbanded in 2015.

A company could not change its market share when it received cars based on the previous year's market share. If you handled 25% of the grain, you received 25% of the cars, which gave you enough space to once again buy 25% of the grain, which gave you 25% of the cars, etcetera—a self-fulfilling prophecy! There were no incentives for improvement or rewards for innovation.

The Pools were the big winners under this system, as the CWB car allocation system preserved their high market share. Other grain companies were fearful of criticizing the powerful CWB, knowing they might not receive their expected share of railcars for grain. Or suddenly, railcars would no longer be directed to their grain terminal at port position. Or incoming vessels would be forwarded to Pool terminals rather than your own for loading and transport to export markets. Control over the transportation logistics was a very complex exercise that was not transparent. Nobody ever really knew what was going on behind the closed doors of the CWB. All participants were afraid of retribution from the CWB, while the Pools were the obvious favourites.

The Prairie grain industry had become political. The alliance between the CWB and the Pools became a political tool as much as an economic tool. Mac Runciman from UGG found this world frustrating.

> *First it is very ideological—in the worst sense of the word, meaning that facts often carry less weight than belief and presupposition. Second, it is a highly political world—again in the worst kind of way, where intrigue, self-interest, and power often play a more important role than reasoned debate.*[18]

At this time, most Prairie towns had a row of line elevators along the railway track. Many were quite small and usually had only a two-to-five railcar spot. The photo below shows the seven grain elevators and one seed-cleaning plant on an elevator row in Mundare, Alberta, in the late 1970s.

---

18   Earl, *Mac Runciman*, page 129.

Courtesy of Len Siracky collection

Left to right:

- Mundare Seed-Cleaning Plant
- Alberta Wheat Pool (formerly Searle Grain, then Federal Grain)
- Alberta Wheat Pool (formerly Searle Grain, then Federal Grain)
- Cargill (formerly National Grain)
- Alberta Wheat Pool (formerly Alberta Pacific, then Federal Grain)
- United Grain Growers
- United Grain Growers
- Alberta Wheat Pool (formerly Alberta Pacific, then Federal Grain)

A major consolidation of line elevator grain companies was occurring during this period. There was little profit in simply handling grain, as the CWB controlled everything and the Canadian Grain Commission set handling fees and storage rates. Many smaller firms sold out to larger firms or merged. One of the largest mergers occurred when Federal Grain took over Searle Grain and Alberta Pacific Grain in 1967. Then the Prairie Pools bought Federal Grain in 1972. Meanwhile, UGG had purchased Gillespie Grain, Reliance Grain, and McCabe Grain. Cargill entered the Western Canadian line elevator business with the purchase of National Grain in 1974.

THE SYSTEM STARTS TO COLLAPSE

What was not widely understood in 1961 was the way a tightly woven web of institutional constraints, arising from well-intentioned but questionable policies, combined to frustrate the development of a modern and efficient system. Nor was there a willingness to acknowledge the role of railway freight rates on grain— the Crowsnest Pass Rates—in the arrested development of the system.

Mac Runciman, President of UGG, 1961–1981[19]

The consolidation of grain companies operating in the country advanced rapidly in the 1950s and '60s. In 1950, there were twenty-three companies with grain elevators. By 1972, after the purchase of Federal Grain by the Pools, there were only eight main players in the country elevator business. In order of size the big six grain companies were: Saskatchewan Wheat Pool, Alberta Wheat Pool, United Grain Growers, Richardson Pioneer, Manitoba Pool Elevators, and Cargill. Two minor players were Parrish & Heimbecker and Patterson, who were smaller, family-owned, private grain companies. The Pools had over 65% of the business in each of their respective provinces.

By the 1970s, the entire country elevator system was old, run-down, and outdated. Most elevators had a railcar spot of three or less cars. The grain companies had not been investing in larger country

19   Earl, *Mac Runciman*, Prefix ix.

grain terminal facilities mainly due to the CWB formula–driven car allocation system. It was futile to build a large, modern facility with a larger and more efficient car spot if you could not get cars for the facility as allocated by the CWB. The money-losing Crow Rate did not give the railways any incentive to invest in more efficient system components since the only tariff they could charge was well below their cost. Also, regulations did not allow the railways to charge a lessor rate/tonne for transporting grain from an efficient facility with a large car spot versus an antiquated old facility with a one car spot. They could not vary the rate structure to encourage more efficient measures. The single-car rate was the only rate. Any new elevator built with a more efficient and less costly multi-car spot could not negotiate a lower rail rate and pass the savings along to farmers to compensate for a longer truck haul from the farm.

Over time, the railroads had not been investing in locomotives and hopper cars for grain movement as the statutory Crow Rate resulted in them losing money on every bushel shipped. Their fleet of old boxcars was being supplemented by new hopper cars being purchased by governments. There were also thousands of miles of high-maintenance-cost branch lines with very little traffic. The railways could not abandon these lines due to the statutory regulations of the Railroad Act and, in fact, were obligated to supply and pick up cars on a regular basis—even at a loss in operations. In many instances, the town had already died, the school and most of the businesses were gone, and yet there was still an old, dilapidated grain elevator open, most often owned by a Pool. As a local church finally shut its doors, one farmer quipped, "Even God has left town, but the elevator is still open!" This system, which was built for the 1920s when grain was hauled by horse and wagon, still existed. Federal regulation stood in the way of any logical, economically driven rationalization.

The grain companies were little more than warehousemen for the CWB. Until canola became popular in the late 1970s, almost all the grain going through the system was for the account of the CWB. Grain company revenue was restricted to handling and storage tariffs regulated by the Canadian Grain Commission. The CWB's policy of holding grain off the international market and paying grain companies to store it was an incentive to keep every old facility open, wherever it was located. These old, crumbling, deteriorating elevators on run-down, barely operable branch lines would be filled with grain and left unattended while the company collected storage fees from the CWB.

This was the system in the 1970s. The CWB and its Pool supporters had reached the pinnacle of their power, authority, influence, and control. The highly regulated, command-and-control grain-handling-and-transportation system; the non-transparent and non-market-responsive operational performance of the CWB; and the Crow Rate disincentive to railway performance in hauling grain all resulted in a system totally dysfunctional for farmers. It was a problem that no number of committee meetings could resolve. Canada was losing its reputation as a reliable exporter of grain as the system failed over and over. The world market for grain was growing and western Canada was missing out. Free market advocates were starting to shout, "The emperor has no clothes!"

By the 1970s, the system had become hopelessly entangled in regulations and bureaucratic zealousness. No one was investing private money into grain elevators or rail improvements. Many farmers had become comfortable with these equality-based dynamics and were fearful of returning to a market-based system even though it was increasingly evident that was just what was needed. Many Western farmers still favoured the CWB and the Wheat Pools. Support for the CWB was strongest among older and smaller farmers.

One organization that was daring enough to start publicly challenging the wisdom of this highly centralized and controlled system was United Grain Growers. UGG was led by a capable, respected, and articulate president, Mac Runciman. As a farmer-owned company, UGG was supportive of open-market solutions rather than the command-and-control practices of the current regulatory agencies involved in grain marketing and transportation.

Much to the chagrin of the Pools that advocated opposite views, the Pools could not say they were the only voice of farmers. Farmers who were members of UGG had an opposing philosophy—that of freedom for farmers.

TRAPPED LIKE THE WILD HOGS OF HORSESHOE BEND!

Years ago there lived a herd of wild hogs in a great horseshoe bend down a river deep in the southern United States. Where those hogs came from no one knew. But they survived floods, fires, freezes, droughts and hunters. They were so wild the greatest compliment a man could pay to a dog was to say it had fought the hogs in Horseshoe Bend and returned alive. Occasionally a pig was killed either by dogs or a gun - and became a conversation piece for years.

One day, a lean-faced man came by the country store on the river road and asked the whereabouts of these wild hogs. He drove a one-horse wagon, had an axe, some blankets and a lantern, a pile of corn and a single-barrelled shotgun. He was a slender, slow-moving man who chewed his tobacco deliberately and spat very seldom.

Several months later he came back to the store and asked for help to bring those wild pigs out of the swamp. He said he had them all in a pen.

Bewildered farmers, dubious hunters and storekeepers all gathered in the heart of Horseshoe Bend to view the captive hogs.

"It's all very simple," said the patient lean-faced man. "First, I put out some corn for them. For three weeks they wouldn't eat it. Then some of the young ones grabbed a cob and ran off into the bush. Soon, they were all eating corn. Then I commenced building a pen around the corn, just a little higher every day. When I noticed they had stopped grubbing for acorns and roots and were all waiting for me to bring the corn, I built the trap door.

"Naturally they raised quite a ruckus when they seen they was trapped. But I can pen any animal on the face of the earth if I can just get him to depend on me for a handout."

Author unknown

In *Canada's Great Grain Robbery*, Don Baron says Runciman would open his talks with the above parable. The message was that anyone can be controlled if you make yourself dependent on them. Prairie farmers had allowed themselves to be controlled by the CWB and government grain bureaucracy. They felt they were getting a benefit from the Crow Rate, but instead it was impeding the system.[20]

Inefficiencies abounded throughout the supply chain. At one time, under CWB management, grain that was originally shipped by railroad to Thunder Bay was then reloaded on railcars and shipped back across the Prairies to Vancouver where it was needed to fill a vessel. Vessels loading in Vancouver would have to be moved and re-anchored several times, as the CWB would have the necessary grain scattered over several terminals. The CWB car allocation system forced the railroads to spot one or two cars at multiple elevators, spread out among multiple delivery points. Each trainload was made up of various types of grain. These trains were then pulled to the north and south sides of the shores of a crowded Burrard Inlet in Vancouver and then disassembled, with a few cars of various products shuttled to the multiple export terminals.

The costs of all these inefficiencies were either hidden in the CWB pool accounts, demurrage and dispatch agreements, or absorbed by the railways under the Crow Rate. There was no transparency in the CWB books that could expose these unnecessary costs. No one was allowed to scrutinize or audit their accounts. Everything they did was secret and hidden from public scrutiny. The Pools and the CWB protected each other in this vicious circle of regulatory self-interest.

The system "hit the wall" in the period from 1967 to 1971. The unravelling started with good crops resulting in large grain surpluses in North America. The CWB thought that if they withheld wheat from the world market the price would go up. By now Canada's share of the world wheat market was down to 26%, while it had been over 40% through the Depression, WW II, and the postwar years. The CWB reduced sales in the period of 1968 to 1970. Sales in those three years were only 63% of the four-year average from

---

20   Baron, Don. *Canada's Great Grain Robbery*. Don Baron Communications 1998. Page 1.

1964 to 1967. Canadian wheat stocks were high. This decision to limit sales, made by the CWB from their office desks in Winnipeg, showed little understanding of the financial strain it was causing Prairie farmers.

At this time, the major wheat exporters in the world were Western Europe, the United States, Canada, Australia, and Argentina, and they all had sufficient supplies of wheat for export.

Hubert Esquirol, a farmer from Edam, Saskatchewan whom I met and came to know through the Western Canadian Wheat Growers Association, related his experience of starting to farm in 1967. Hubert still has his CWB delivery-permit books for the crop years 1967/1968 through to 1970/1971. In that four-year period, his allowable delivery quotas were 6 bushels/acre in 1967/1968, 5 bushels/acre in 1968/1969, 4 bushels/acre in 1969/1970 and 8 bushels/acre in 1970/1971. At that time and in that area, wheat yields were about 40 bushels per acre. How was a farmer supposed to survive financially when over those four years only 15% of production could be sold? Farmers were forced to summer fallow to get more quota acres on which to apply wheat deliveries. (See Hubert's CWB permit book photos on pages 68 -73.)

In the 1960s Jack Gorr graduated from the University of Alberta with an Agriculture degree. He joined his father and brother on their family farm near Trochu, Alberta. With the restrictive CWB delivery quotas in the late 1960s and early 1970s, Jack was forced to leave the farm and seek full-time employment elsewhere. Later, when grain prices increased and grain delivery restrictions eased, Jack returned to the family farm and made it his lifelong career.

The CWB appeared oblivious to this hardship in the country. Prairie farmers had no options or choices but to live with CWB decisions. Others survived by diversifying into livestock production. Many were forced to seek off-farm income to offset the lack of grain-sales opportunities.

The late 1960s found Prairie wheat farmers desperate for cash flow. Wheat was stocked up everywhere. Mixed farms were doing somewhat better, as they had livestock to sell for income.

Saskatchewan grain farmers began desperately and illegally shipping barley to Alberta cattle feeders for prices as low as three bushels for a dollar. At the time, interprovincial movement of grain, even feed grain, was illegal and controlled by the CWB. Fines and penalties were imposed on farmers who were found to be hauling grain across provincial boundaries.

The livestock industry in the West was disadvantaged by freight subsidies on feed grain to central Canadian livestock producers. Livestock production on the Prairies was further hindered by the Crow Rate, which provided a subsidy for shipping grain off the Prairies rather than processing or feeding it here. However, in practice, livestock production in the West was often the only way farmers could market their grain—by feeding it. The CWB was not aggressively selling in the world market, adhering to the ill-advised and timeworn concept that if they just held grain off the market the price would go up. By now Canada was a small player in the world market and withholding supply had little effect on world price. Farmers suffered the consequences since there was little cash flow from grain for several years.

In 1971 the federal government, through Wheat Board Minister Otto Lang, brought in the Lower Inventories for Tomorrow (LIFT) program. Farmers were paid six dollars an acre to idle their land and stop producing wheat to slow the growth of wheat stocks. The spike in summer fallow acres that year is clear in the preceding graph on Prairie summer fallow acres. (page 76) The LIFT program served its purpose and achieved its objective, but at the wrong time.

On the other side of the globe, hidden from world knowledge, a large, grain-producing entity was becoming desperate to purchase wheat.

In 1972, an event that came to be known as the Soviet Great Grain Robbery occurred.[21] Unknown to the rest of the world at the time, the Soviet Union was short of wheat and having trouble feeding

---

21 This escapade from Cargill's perspective is well explained in the historical fiction account of the event by Crawford, Russ. *Limit Up*. Agrinomics Publishing. 2022.

itself. In 1971, and particularly in 1972, the Soviets faced short crops due to droughts. They desperately needed to purchase large amounts of grain internationally. World prices had been languishing for some time as surplus stocks, primarily in North America, were large.

Starting in July of 1972, the Soviets sent a buying delegation to North America to meet individually and secretly with various wheat exporters from the US and the CWB from Canada. Before the market realized what was happening, the Soviets had purchased most of the surplus grain in North America at low prices and without alerting the market. Many of the American sales included export subsidies from the United States Department of Agriculture (USDA). Only after all these deals were completed did the industry fully realize what had happened.

Following this huge transaction, the price of Chicago wheat more than tripled in fifteen months from $1.43/bushel in June of 1972 to $5.32/bushel in September of 1973. American supplies of soybeans became so low that an embargo was placed on further exports to preserve sufficient supply for domestic use.

This momentous event changed the way the world traded grain. Instead of the world seeing the Soviets as exporters, they had now become importers. As grain stocks diminished and the price of grain skyrocketed, both land prices and interest rates shot upward. The Alberta economy was booming in the 1970s thanks to the rising price of oil and the firm leadership of the Lougheed Progressive Conservative government. But farmers were missing out on the boom.

Largely due to the federal government's Lower Inventories for Tomorrow (LIFT) program in 1971, Western Canadian acres of summer fallow increased while acres sown to wheat decreased. With this policy of discouraging wheat production in 1971, Western Canada found itself with reduced stocks of wheat in 1972. However, the world now wanted more grain, and prices were sharply higher. In hindsight, LIFT was a financial disaster as the reduced planted acreage resulted in Canada having little grain to sell in the new environment of higher prices.

After the Soviet purchases, both grain production and trade became quite profitable. Canadian production increased and Canada was able to make lucrative grain sales in a growing and profitable world market. But we could not get our grain to port for our customers to load on a vessel. In addition to the institutional impediments as described above, the system was plagued with strikes by railway and port workers and other unions along the supply chain.

There was a diminishing fleet of old and worn-out boxcars that were insufficient for the sales being made. The railways were hesitant to invest in locomotives and grain cars as they were losing money under the Crow Rate. The CWB, Pools, and leftist National Farmers Union (NFU) were opposed to changing the Crow Rate and having farmers pay more for shipping grain. They argued the rate for grain transport should not be changed since the CPR had earlier received land and mineral rights concessions from the government.

It was a standoff, and everyone was a loser when a grain sale could not be completed.

≈

During this frustrating time in the 1970s when grain sales by the CWB were sluggish and transportation problems abounded, a savvy group of Southern Saskatchewan farmers and agribusiness men banded together and began questioning why the system was not working. This led to the formation of the Palliser Wheat Growers Association, later renamed the Western Canadian Wheat Growers Association (WCWGA). They, along with the Western Barley Growers Association, become the pre-eminent farm groups in Western Canada promoting market-based solutions. The Wheat Growers became an irritant to the CWB, which was not accustomed to this type of resistance and questioning from organized farmers. This group elevated the issue of wheat marketing to a higher level of discussion, as more and more farmers began to question the operational success of the regulated marketing-and-transportation system.

In 1972 members of the then Palliser Wheat Growers organized a unit train shipment of hopper cars of wheat to Vancouver. It was loaded at the government terminal in Saskatoon. One train, loaded at one location with one product, was pulled as a unit to Vancouver. This proved it could be done. But regulations that prevented it from becoming commonplace would not be changed for another forty years.

This feat was not easily achieved. Since the CWB oversaw grain elevators, boxcar allocation, and assignment of Vancouver terminals the cars would unload at, some manner of getting around the CWB regulations had to be found. The CWB was asked to cooperate and try this experiment, but they would have no part of it. The canny Palliser group found an answer in discussions with Federal Grain. Federal owned the Neptune bulk-loading facility in Vancouver. They also owned hopper cars they used for potash that could be used for wheat. The wheat was cleaned, inspected, and made ready for export before it left Saskatoon. CN spotted a unit train of cars at Saskatoon and pulled them to Neptune in Vancouver. Here the wheat was loaded directly onto a vessel. It could be done!

Federal Grain was so proud of the achievement that they had the entire project filmed. Several months later, SWP purchased Federal Grain and the film disappeared.[22] Wheat Growers members recovered the film and played it at a Wheat Growers annual meeting years later, after SWP went public and became Viterra. Until Harper's changes to the CWB Act in 2011, it would take another forty years for the system to be sufficiently deregulated to have unit trains become the norm in Prairie grain shipment.

In 1974, after demands from the Quebec livestock industry, feed grain for livestock was removed from the CWB mandate and was allowed to trade on the open market. Interprovincial trade in feed grains was legalized and domestic feed grain could then move outside of CWB control. In 1979, the CWB, citing the need to control the flow of domestic feed grain into the licensed system, persuaded Ottawa to allow them to place delivery quotas on

22 Baron, *Canada's Great Grain Robbery*, Pages 148–150.

domestic feed grains. All provincial agriculture ministers (except the Saskatchewan NDP minister) opposed this measure.

In the early 1970s, the government formed the Grains Group in Ottawa, a small unit of intra-departmental grain specialists who reported to the CWB Minister. The mandate of the Grains Group was to advise the Minister on how to promote and improve grain movement. It was here, working for the Grains Group in Ottawa in the early 1970s, that Paul Earl started his long career in policy development for the Prairie grain industry. Over time, many positions in the Grains Group became filled with former employees of the CWB. Individually they became CWB advocates in Ottawa's political and bureaucratic scene, and not independent civil servants. Though this was not the formal policy of the Grains Group, this undercurrent of unwavering CWB support permeated the group. I witnessed this personally when I was there in 1979–1980.

Increasingly, the private grain trade and industry participants along with progressive farm groups were becoming critical of the way the system was being controlled by the CWB. By now the CWB had become an untouchable and powerful bureaucracy. The three Prairie Wheat Pools and their farmer supporters provided a strong body of unwavering support for the current system of marketing and controlling grain flow. They called it "orderly marketing," a term with an ambiguous meaning. Even politicians and senior government officials in Ottawa were hesitant to question the CWB. The Pools worked closely with the CWB.

At the encouragement of the federal government, in 1969, the industry formed the Canada Grains Council. It included a wide range of senior industry representatives. Members included the grain companies, the railroads, farm groups, and other industry participants. Their purpose was to find ways for all parties to act together and get grain moving. However, political schisms within the members of the Grains Council, along with opposing and uncompromising views of what a new system should look like, rendered the existence of the CGC totally ineffective. This was another failed move to reform the system.

As the transportation-and-handling system was breaking down, in 1973, the Grains Council studied the situation and concluded major changes were necessary. The study described the grain-handling-and-transportation system as "an industry in shackles, a massive bureaucracy unable to respond to beckoning markets."[23] The Pools and NFU left the Grains Council after the Council accepted this report. The CWB never wanted to have anything to do with the CGC from the day it was formed. The influence of the Pools and CWB totally overrode the warning signs of coming capacity concerns cited in the study, as sufficient capital was not being invested by the railroads and grain companies to modernize and expand the system.

The arrogance of the CWB at this time is best reflected by comments made by CWB Chief Commissioner Esmond Jarvis in the late 1970s. Jarvis, a lifetime bureaucrat, went so far as to publicly chide the railroads and grain companies for not spending more to modernize the system. At the same time, the CWB had full control over the deployment of these assets.

Some participants in the grain industry were starting to question the economic wisdom of the statutory Crow Rate. The Pools, left-wing farm groups, and the CWB supported maintaining the fixed rate, even though it was clear the system could not work when the railways were losing so much money.

The government also established the Senior Grain Transportation Committee (SGTC), which included senior members from across the industry, including some elected farmers. They met regularly at the senior executive level (rather than operational) to see if some resolution could be found on the impasse in system improvements. The formation of all these organizations, and the numerous meetings that followed, was simply like "shuffling the deck chairs on the Titanic." The real iceberg to the system was the Crow Rate and CWB domination with their uncompromising and clandestine style of system management.

---

23   Baron, *Canada's Great Grain Robbery*, pages 194–95.

One of the only constructive outcomes of all this study was the government purchase of hopper cars to replace the aging boxcar fleet. The railways hadn't purchased any new boxcars for grain since the 1950s, and by 1973 the number of boxcars available for hauling grain was half of what it was a decade before. [24]

---

24 Baron, *Canada's Great Grain Robbery*, page 157.

The above photo is myself on the ladder of new Government of Canada hopper cars spotted at UGG elevators in Mundare in the mid 1980s. In the 1980s and 1990s I spoke at many farm meetings advocating for a reformed and more efficient transportation system for grain. Note the top hatches on the car are open for loading, and the hoppers at the bottom of the car are where the grain was discharged. This was a tremendous advance from the small, awkward and inefficient boxcars.

The capacity of a hopper-car was greater than that of a boxcar, and they were much easier to load and unload. This was but a small bandage for the gaping wounds in the total system. Starting in the mid 1970s, and over the next while, the federal government and the Provinces of Saskatchewan and Alberta, would purchase hopper cars, adding 15,500 hopper cars to the grain fleet while governments and industry and farm groups debated what to do about the Crow Rate. Even the CWB purchased 4000 hopper cars taking farmers' money from the pool accounts to do so.[25]

Ottawa also addressed the financial deficiency gap in rail rates with several ad hoc, band-aid programs, many of which were more politically motivated than practically effective. Hundreds of millions of taxpayer dollars were spent on rehabilitating old boxcars and branch lines and subsidizing the railroads to keep the branch lines open. These funds were a total waste of money as the boxcars and branch lines were soon to be abandoned anyway. The strongest advocate of these expenditures was SWP, who had most of the small, old, crumbling elevators on these branch lines, full of grain that was collecting storage fees from the CWB.

If a businessman makes a mistake, he suffers the consequences.

If a bureaucrat makes a mistake, you suffer the consequences.

–Ayn Rand

---

25 A good account of the government purchases of hopper cars can be found on the web site: Churcher, Colin. *Requiem for Government Grain Hopper Cars.* Colin Churcher's Railway Pages. July 30, 2021.

KING WHEAT MEETS THE CINDERELLA CROP

Rapeseed (now canola) acreage was increasing, and it was becoming a prominent crop. Rapeseed was a European crop and had a high acid content. Canadian plant breeders went to work and were successful in creating low-acid varieties, which were refined into healthy cooking oils. To distinguish high-acid rapeseed from these new varieties, the crop was renamed canola (**Can**adian **O**il **L**ow **A**cid). Also, late in 1973, Prairie growers won a plebiscite to keep rapeseed marketing out of the control of the CWB, with 78.5% of farmers voting in favour. The Pools and NFU had campaigned unsuccessfully to have rapeseed marketing placed under the control of the CWB.

After its hundred-year domination of the Western grain industry, the supremacy of wheat was about to be challenged. Canola acreage was increasing as the prairies were agronomically favourable to its production. Over the next fifty years canola production would grow to match, and eventually exceed, that of wheat. It was dubbed the Cinderella Crop. Canola was not marketed through the CWB. As canola and other new crops were introduced to the Prairies, the domination of the CWB over the grain industry began to fade. Increasing acreage of other new crops included lentils, dry field peas, canary seed, mustard seed, and chickpeas. These new crops were thriving on the Prairies, and they were all being sold by farmers on the open market directly to competing grain companies.

Following is a graph showing farm cash receipts from wheat and canola production over the last fifty years. After one hundred years of dominance on the prairies, King Wheat was being replaced by the Cinderella Crop.

Farm cash receipts from canola and wheat in billions of dollars

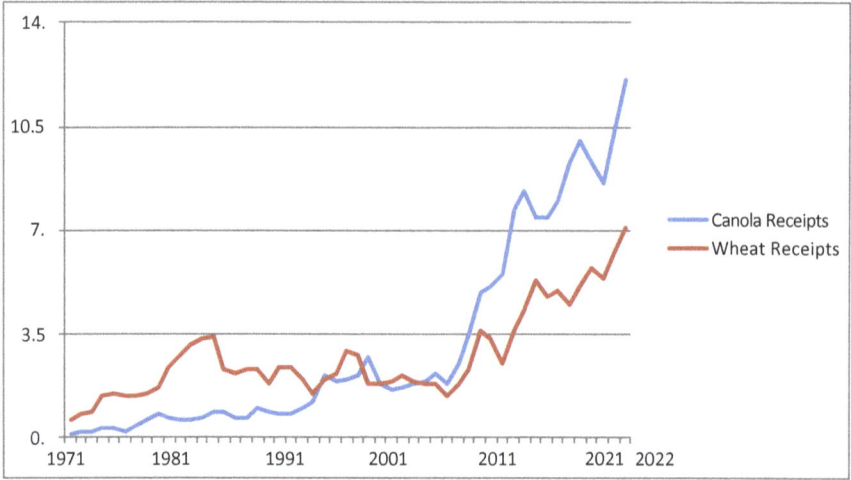

≈

Courtesy Marlene Stefanyk Bodnar. September 1977.

This is what a typical, aging prairie-grain elevator looked like in the 1970s. The grain-handling-and-transportation system still revolved around the metrics of the 1920s, with many small facilities with small railcar spots scattered along many miles of little used railway branch lines.

This is a 1977 photo of an Alberta Wheat Pool elevator at Hilliard, Alberta, located west of Mundare. It was managed for many years by Bill Stefanyk. There are several portrayals in this photo. Firstly, all the railcars are still boxcars. The main tall facility housed the "leg," which was a bucket elevation system that took the grain up and then discharged it to flow downward in the drop spouts to be stored in the various bins and annexes. The main facility would usually have several smaller bins.

To the left of the main house was the crib annex, designed as a permanent structure. It was the main storage facility and was sturdily built with 2x4s or 2x6s lying flat, one on top of the other. The crib annex was connected to the main house by a drop spout and then a drag auger on top moved the grain to the various bins. There was also a drag auger at the bottom of the crib annex that would move the grain back into the main house to turn grain or prepare it for shipping.

To the right of the main house was the balloon annex, designed as a temporary storage facility. In the early years of World War II, and then again in the late 1960s when there were large stocks of wheat, these octagonal annexes were quickly built in response to storage payments by the CWB and the Federal Government. They were not of very sturdy design and were given the rather unflattering name "balloon" due to their sidewall's tendency to sag and balloon outward when filled with wheat. The facility in this photo was connected to the main workhouse by downspout for filling, and by dragline at the bottom for unloading.

On the far right behind the warehouse was yet another balloon annex. This annex was likely constructed last, when there was no longer room to build on either side of the main elevator. Once again, grain could be elevated up the main leg and deposited by drop spout into this annex. The problem with this annex was that it was not connected to the main facility by drag auger. To clean it out, the grain had to be loaded onto a truck and hauled back to the main leg. This was a labor-intensive process, and this annex was used for long-term storage rather than "quick turns" of grain. These balloon annexes were usually just one large bin and not divided into small bins like a crib annex.

To prevent the walls from bulging when filled, a series of wires crisscrossed the inside of the balloon annex, connected opposing walls to hold the structure intact. All these wires and no dragline made it a daunting task to unload the grain from this facility. I experienced this myself when working for our Searle elevator manager, Bill Fedoruk, in Mundare in the 1960s, shoveling grain amidst the seemingly endless crosswires.

Lastly on the right foreground is the storage warehouse. Here, eighty-pound bags of fertilizer could be unloaded by hand from boxcars. This was before the days of bulk fertilizer handling. I also witnessed this happen, but fortunately was not old or strong enough to handle the bags at the time. Also stored in this warehouse would be herbicides, twine, fencing supplies, and other goods for farm use.

BACK AT THE FARM IN MUNDARE AND ON TO OTTAWA

The thousands of miles of wilderness between East and West has a political significance that should not be overlooked.

–Tom Crerar, President of UGG, 1907–1929 [26]

This was the situation on the grain-producing prairies when I seeded my first crop in 1977. I was working in Edmonton at the statistics branch of Alberta Agriculture. I always wanted to farm, and for the 1977 crop I convinced my father to discontinue summer fallowing and begin using nitrogen fertilizer and more herbicides to control the growth of wild oats. It was also the first year I seeded my own crop. Canola was a relatively new crop at the time and seemed quite profitable. Besides being profitable, one of the biggest benefits of producing canola was that it was marketed by grain companies and not the CWB, and it provided quick cash flow after harvest.

That year our farm seeded about 550 acres, all to wheat and canola. Wheat sales off the farm were restricted to the usual 3 bushels/acre opening quota. Even though the CWB did not market canola, they imposed a 3 bushels/acre opening quota on canola deliveries into the elevator system with their logic that "the CWB didn't want canola farmers plugging the system and restricting the ability for others to

---

26   Rae, J.E. *T.A. Crerar A Political Life.* McGill–Queen's University Press. 1997. Page 114.

deliver wheat." With the higher cash outlay for additional inputs, we needed cash flow soon after harvest.

I found out that there was a canola-crushing plant in Lethbridge by the name of Canbra Foods that would come right to your farm and pick up and haul the canola to Lethbridge by truck at their cost. This was a vast difference from the restrictions we had to cope with selling wheat. There was an opening quota of 3 bushels/acre for wheat and we had to haul the grain to the elevator in Mundare. Canola to crushers had a 10 bushels/acre quota and Canbra covered the cost of trucking from our farm to Lethbridge.

Even though the CWB did not market canola, and canola going to Canbra Foods in no way affected deliveries of wheat, the CWB still imposed an opening quota of 10 bushels/acre to crush plants. They cited the reason was to "allow equal delivery by all farmers and not have just a few delivering all their grain while others waited"

As the year went on, the CWB would slowly increase the delivery quotas for all crops as export sales were made and additional elevator space became available. But the restrictive opening quotas still made it difficult to contract and deliver grain off the combine for quick cash flow. Since we had used more fertilizer and herbicide that year, our need to deliver grain and obtain some cash was greater than that of most farmers who were still summer fallowing. The economics of the higher inputs worked since our crop yields were much better. But we faced the roadblock of delivery quotas, which were the same for everyone. Some farmers' cash flow needs are different from others, but the CWB felt they had to ensure everyone had an equal chance to deliver their opening quota before they would open another.

Delivery quotas were not quick to open at this time since the marketing-and-transportation system was so backlogged due to the problems previously described.

This was the grain-marketing environment in 1979 when the Conservative Joe Clark minority government was elected in May of that year. Don Mazankowski, our local MP, was named minister of transport with responsibility for the Canadian Wheat Board. I knew Don Mazankowski (Maz) as he was our MP, and I had been a junior member of his constituency association.

I was the youth delegate from the Vegreville constituency at the leadership convention in Ottawa in 1976 where Joe Clark was elected leader. That was my first plane ride, and I travelled with John Batiuk, our neighbour and local Alberta Member of the Legislative Assembly (MLA) in the provincial Lougheed Conservative Government, and my uncle, Peter Polischuk, who was John's campaign manager and very active in Conservative politics. I had my first taste of Ottawa.

In July of 1979 I wrote Maz, the new minister, a long letter[27] describing my frustration in trying to market grain under the current untenable situation. The bottom line of the letter was that until two changes were made, our industry could not flourish on the Prairies. The two necessary changes needed were:

- Diminish control of the CWB and let farmers market their own grain.
- Start compensating the railways for hauling grain.

A couple of weeks later, I was contacted by Peter Thomson, Maz's senior policy advisor and political confidant, asking to interview me. Peter flew to Edmonton, and I was interviewed by him at the Macdonald Hotel in July of 1979. He offered me a job working in Maz's ministerial office in Ottawa. I was ecstatic and said I would likely take the offer, though I first wanted to discuss it with my wife, Wendy. It was a complex time in our life as Wendy was expecting our first child in September. I also felt a responsibility to assist with the harvest as I was helping my parents on the farm, and I had some crop of my own seeded. I was also torn between leaving the farm,

27   A copy of the original letter is listed as Appendix 2.

since that is always what I wanted to do, and going to Ottawa to participate in the excitement of a new government. We decided this opportunity was too good to pass on and we decided to go to Ottawa for a couple of years, likely the period that the minority government would hold. The farm would wait until we came back. We agreed I would go to Ottawa that November.

Since the government was in a minority position, Maz was mindful of my job security. Most of his staff were of politically exempt status, meaning if he went, they went. In my situation, Maz arranged for an executive exchange program between Alberta Agriculture and the Grains Group in the federal Department of Industry, Trade and Commerce. This was a similar arrangement that assistant Don Adnam had had with the former Liberal Minister, Otto Lang. So, a precedent had been set. It was to be an official two-year program where I was to continue to be paid by Alberta Agriculture but would be posted with the Grains Group in Ottawa. Maz then seconded me to his office, and my employment was secure.

I left Alberta that Remembrance Day weekend and went to Ottawa alone, leaving my family at home on the farm for now. It was a big leap for a Prairie boy who had never been far outside the Mundare/Edmonton bubble. It was difficult for me to go alone to a totally new experience.

Arriving in Ottawa was a surreal experience. Legions of bureaucrats with bulging briefcases and exaggerated looks of self-importance trudged through the slush. French was spoken everywhere, and since I did not speak French, I felt like I was a foreigner in my own country. If I mentioned I was from the West, I was treated as some troublesome, odd specimen to be gawked at. The bureaucracy was almost all from Central and Eastern Canada since bilingualism was mandatory in the civil service. Knowledge and ability seemed secondary. Ottawa was a lot further from the West than just distance would suggest.

Things hadn't changed much in sixty years. In 1917, UGG President Tom Crerar was appointed minister of agriculture in the Borden Unionist government. After arriving in Ottawa, he wrote to a friend, "[I]t was like entering a museum . . . the whole atmosphere of the place tended to stifle or depress initiative rather than encourage it."[28]

As I entered the hectic turmoil of the minister's office, I tried to find out where I fit. I soon found the job of a ministerial assistant was not to influence change, as I had hoped. Rather, it was to assist the minister in furthering the agenda of the government, provide liaison with departmental staff, and cooperate with all other ministers' staffs on a political basis. While trying to learn what I could do to contribute, the ill-fated bungling of the Joe Clark government resulted in the loss of a non-confidence vote in December of 1979, whereby the government was defeated.

Here I was planning to move our family to Ottawa after Christmas, and an election was called for February of 1980. Ottawa essentially goes into limbo during an election, with the only activity being the administration of ongoing programs. The House and Senate do not sit. There are no new initiatives being planned or implemented, and the politicians are home campaigning in their ridings.

The Clark Conservatives went on to lose the February 1980 election.

Overnight Maz and the Conservatives were out, the ministerial staff were out, and another Pierre Trudeau Liberal government was elected. Since I was on an exchange program with the Alberta government, I still had my job, but no office, no responsibilities, and little to do. I had to vacate my office, which was taken over by the staff of Jean Luc Pepin, the new transport minister.

Trudeau had few elected members from the West to choose from to appoint to a cabinet position. Rather than an elected Member of Parliament becoming minister, Trudeau appointed Liberal Saskatchewan Senator Hazen Argue as minister responsible for the Canadian Wheat Board. Argue was the former leader of the Canadian Co-operative Federation (CCF), forerunner to the

---

28   Rae, *T.A. Crerar A Political Life*, page 48.

current New Democratic Party (NDP). Argue was a staunch CWB supporter.[29]

Bill Miner, head of the Grains Group, gave me an office with a desk and a phone, but few, if any, responsibilities. I was a political leper, and nobody knew what to do with me. The first task I was given was to write some briefing notes for Minister Argue. I completed my assignment, forwarded it up the system, and never saw it again. Subsequently, I was never requested to prepare any more briefing notes. Because of the executive exchange program I was on, I was still technically employed by Alberta Agriculture, so everybody simply ignored me, not caring whether I came to the office or not. But I did come in every two weeks to pick up my pay cheque! Here we are in Ottawa, my family and I, locked into a two-year contract with essentially nothing to do.

In the spring of 1980, we spent time travelling around and touring Ottawa, Eastern Canada, and the Maritimes. I spent time in Maz's MP office two or three times a week with former ministerial staffers Jamie Burns (son of James Burns, CEO of Power Corporation), Pat Walsh (future government liaison for Irving Oil), and Bill Elliot (later RCMP Commissioner). We would meet about 10:00 a.m. and play cards until early afternoon, taking a break for a subsidized lunch at the House of Commons's subsidized cafeteria— often a toasted club sandwich and a glass of milk for $1.25. Some weekends we would go to Montreal with Jamie and his wife Terry and stay at his parents' home in Westmount. We toured Montreal in this manner, and in the evenings, we would go out on the town to explore Montreal's nightlife.

As the weather warmed up, we took road trips. We spent Easter in Pennsylvania with Wendy's relatives; toured a bit of Washington, DC; and experienced a New York traffic jam during a subway strike. We spent several days touring the New England

29 Apparently before Argue was appointed, the CWB ministry was offered to Mac Runciman. Pierre Trudeau offered to appoint Runciman to the Senate, and then make him minister. According to the lore, Runciman refused as he believed a minister should be an elected Member of Parliament, not an appointed Senator. As always, Runciman was a man of strong principles.

states, and then some time exploring the Atlantic provinces. Of course, I was still employed and collecting my pay cheque from the Alberta government.

As May rolled around, we went back to Alberta to help put the crop in and then came back to Ottawa homesick. Getting paid for doing nothing substantial may sound appealing but is not intellectually fulfilling. Wendy's maternity leave was expiring at the end of June, and she could still return to her job in Alberta. That sealed it! We were going back to Alberta.

I contacted human resources at Alberta Agriculture and requested to reverse the exchange so we could go back to Alberta. This allowed us to have our moving costs covered by the government. I was told they were not going to allow it—it was not conventional, and it would have a negative effect on the integrity and political neutrality of the program. After a good bit of wrangling, I finally said, "I quit. I will pay my own way back." This was not something they wanted to hear, but they relented, and we returned to Mundare in June of 1980. It did not take me long to clean my desk at the Grains Group. In Alberta, I resumed my job at the statistics branch of Alberta Agriculture and continued to farm, looking to get more land to farm.

GRAIN TRANSPORTATION AUTHORITY (GTA)

When Don Mazankowski became minister of transport with responsibility for the CWB in 1979, he quickly took steps to address the problems plaguing the grain-handling-and-transportation-system. Cataloguing the rabbit warren of rules, regulations, and overlapping bureaucratic jurisdiction was like peeling off layers of an onion. Take one off and there was another just below it. Maz named a three-person committee of Western MPs to investigate these issues. This committee was chaired by Saskatchewan MP Bill McKnight and included Alberta MP Stan Schellenberger and Manitoba MP Jack Murta. Upon review, one of their recommendations was to establish an oversight body in the system to identify and address obstacles. This body was to assume some of the CWB car allocation responsibilities for non-CWB grain. It was named the Grain Transportation Authority (GTA).

Dr. Hugh Horner (Doc) was appointed to head up the organization, a small group that included senior experienced personnel from all sectors of the industry. Doc Horner was previously deputy premier of Alberta, Lougheed's rural lieutenant, and a cabinet minister. He hoped to introduce some constructive changes to the regulated Winnipeg grain establishment. The GTA office was in Winnipeg, but Doc wished to remain in Alberta, so he set up an office in Edmonton where he worked on Mondays and

Fridays. The Edmonton office was essentially a waypoint for his travel back and forth to Winnipeg.

From his previous deputy premier's office, Doc took with him two of his trusted staff, Doug Radke and Tom Burns, to set up the GTA with him. Doug moved to Winnipeg and became the deputy administrator of the GTA, while Tom staffed the Alberta office. When the Clark government fell, and the Trudeau Liberals were in charge again, Doc resigned and Doug Radke remained as acting administrator. Doc did not wish to work for a Liberal government and neither did Tom Burns. The responsibilities of the Alberta office were now largely diminished since Doc was not around to use it.

Prior to the changes Mazankowski initiated at the GTA, the CWB was solely responsible for all railcar allocation for all grain in Western Canada, both board and non-board grain. Now the GTA took over negotiations weekly with the railroads for total car supply. The GTA also took over allocation of producer cars, and cars for shipment of feed grain to the East, which were a priority. As well, the GTA took over allocation of cars for non-board-grain sales made by the private trade. All cars for CWB grain went en bloc to the CWB to distribute as they chose. There was no end to the complaints from the CWB that they were not getting their weekly share of cars, yet they would not share sales information with the GTA to prove what they needed as did the private trade for non-board grain.

Soon after I returned to Alberta in the summer of 1980, Tom Burns called me and informed me that he was leaving the GTA and asked if I would be interested in the job of managing the Alberta office. I immediately accepted—here was a chance to get back into the tangle of the regulated grain-marketing-and-transportation system and to influence positive change. This was the autumn of 1980.

The main work of the GTA was done in Winnipeg, and the Edmonton office had lesser responsibilities. However, it gave me a chance to work in the system and continue to farm at Mundare. We were living on the farm, and I commuted to Edmonton daily. Doug Radke was a capable administrator and was supportive of my situation of not wanting to move to Winnipeg. Like Doc, he also wanted to see changes to the regulated grain industry in Winnipeg. I was welcomed by him as an ally in attempts to unravel the regulations impeding positive change. While at the GTA, I often got myself into trouble advocating for deregulatory reform. There were complaints from the Canadian Wheat Board and Saskatchewan Wheat Pool about my comments. Radke assured me that as long as I was not speaking on behalf of the GTA, I was free to express my personal views.

At the GTA I met co-worker Paul Earl, and came to respect his knowledge, research, and analysis of the industry. Later, Paul and I would also work together at UGG and the Western Canadian Wheat Growers. Paul went on to the world of academia at the University of Manitoba and published several books and papers on the Western grain industry. Many times, through the years I would become disheartened by the lack of progress in deregulating the system. Paul was always there to encourage me to keep going, citing a quote from John Stuart Mill in 1867: "The only thing necessary for the triumph of evil is for good men to do nothing." This helped keep me going.

I spent much of my time working at the Winnipeg office, travelling back and forth from Edmonton. One of my activities was to assist staff member Adrian Measner in his weekly railcar allocation responsibility. Adrian was from the CWB and later became its CEO. He was a staunch supporter of what the CWB advocates called orderly marketing, one organization marketing all farmers' grain. We had fundamental differences of opinion on what the future of the industry should look like, but I learned the formulas used for railcar allocation.

The workload was light in the Alberta office, so I spent my time going to industry meetings, farm meetings, and anywhere I could to learn more about how the pieces fit together in the regulatory puzzle and who the main players were. I met many interesting and accomplished people. I spent the quiet days in the Edmonton office reading and learning. If my travels were linked to grain-handling-and-transportation issues, I had the authority to attend. By the spring of 1984, I was getting bored and restless with so few responsibilities. The GTA needed the position in Winnipeg where they were short-staffed, and I did not want to relocate. It was much more difficult to influence policy change than I thought. I resigned in 1984 and went home to seed the crop.

The farm had been slowly growing every year through additional land purchases and rentals. By 1984 our seeded acres were up to 1,300. In 1982, we built a house and then were permanently established on the farm.

Aerial photo of our house being built in 1982, with the mobile home we lived in from 1978 to 1982 in front of our new house. Our residence was across the road from the main farmyard where I grew up. The property our house is on was homesteaded by my grandfather, Ivan Motiuk, in 1899.

THE BATTLE OVER "THE CROW"

Through all this time, the issue of compensating the railroads for hauling Western grain was overhanging the industry. The status quo was unsustainable. The railroads were losing millions of dollars annually on Western grain. Most parties, except SWP and the far-left NFU now agreed that the old Crow Rate structure should be reviewed. The "Crow Gap" was identified as the difference between what railways were being compensated and what their costs were. It was also agreed by most parties that this Crow Gap should be paid by the federal government since Western farmers could not financially absorb such a large, sudden increase in freight rates. Differences of opinion arose as to the way the subsidy should be paid. It became a controversial, bitter, and divisive debate in the Western farming community.

The Pools and other parties who were resistant to changing the freight rates wanted the government to simply compensate the railways annually for the Crow Gap. They did not wish to see any change that might impact the large grain handle that was their lifeblood. Livestock groups, value-added industry, farm-diversification advocates, right-wing farm groups, and many economists argued that this would preserve the economic irregularities and regulatory inefficiencies in the system. Wrong market signals would be sent to farmers as to what the actual cost of shipping grain was. These groups lobbied for a pay-the-producer type of subsidy whereby farmers would pay the full freight rate and in turn receive an annual subsidy from Ottawa for the Crow Gap. It was argued for years that the Prairie livestock industry was economically disadvantaged due to the artificially low freight rates. This essentially was a subsidy to export grain, thus making feed prices to livestock farmers higher than they should be.

Liberal Transport Minister Jean Luc Pepin hired Winnipeg academic Dr. Clay Gilson to study the entire issue and report back. After broad consultation with interested parties, Gilson released his report in 1982 where he identified the Crow Gap (the difference between what the railways were compensated and what their costs were) as $646 million annually. As for the contentious method of payment, Gilson recommended that initially the payment be made to the railways, and by 1990 a payment to producers be phased in to address the long-term market inequities the fixed rate created. Pepin accepted the report's recommendations and proposed the government proceed with it.

The Pools were staunchly opposed to a payment to producers. They saw it as the thin edge of the wedge that would disrupt the regulated system, on which their large number of small, inefficient elevators relied. They went so far as to lobby Quebec farm groups, convincing them that a payment to producers in the West would negatively affect the livestock industry in Quebec. Quebec farm groups turned on their lobby machine and convinced the Trudeau government that the payment should be made directly to the railroads. The lobby was successful due to an unholy alliance of the three Prairie Pools and Quebec farmers. Politics does indeed make strange bedfellows. The government went ahead and passed the Western Grain Transportation Act (WGTA) in 1983, with a payment of the Crow Gap to be made to the railways.

BACK INTO POLITICS

Early in 1984, after I left the GTA, Don Mazankowski asked me to be his campaign manager for the upcoming federal election that was expected later in the year. I accepted. The election was announced for September 4, and it was an extremely busy summer for me. Maz was popular across the country, and he spent much of his time campaigning outside of the riding. The work in his Vegreville constituency was left to me and the team he put together. Since Maz was so popular at home, recruiting and organizing workers was easy. The same cannot be said about working with others in the core campaign team, which included many of his old-time buddies in Vegreville who were quite set in how they thought things should be done. Two of Maz's main confidants in Vegreville were Bill Yurko and Eugene Demkiw. Bill went on to be an assistant commissioner with the Canadian Grain Commission, and Eugene went on to become a director on the board of CN Rail. Ed Bilyk was Maz's official agent and did a good job ensuring the campaign followed the strict spending rules each candidate had to follow during the election. With his overwhelming popularity Maz won the Vegreville riding with the highest plurality in Canada in the Mulroney sweep of 1984.

1984 Don Mazankowski Campaign Team

Left to right standing: unknown, unknown, Kay McKenzie (returning officer), Ken Motiuk, Ron Rudkowski, Ed Bilyk, Peter Polischuk Jr., Bill Mandiuk, Jackie Yakemitz, Daniel Dubuc, Bill Yurko.

Front and centre: Hon. Don and Lorraine Mazankowski

In the new Mulroney government of 1984, Maz was again appointed minister of transport with responsibility for the CWB. Again, I saw a possibility to become involved in influencing the deregulation of the grain industry. I was hampered by the fact that I did not wish to move to Ottawa to work in his office there. By now our family was well established on the farm and we decided it was to be our long-term livelihood. We had to make a pivotal decision for the future—move to Ottawa and stay involved in politics and influence change from the top down or stay in Alberta, continue to grow the farm, and try to influence change from the bottom up. We chose the latter.

I was offered the job of managing Maz's Vegreville constituency office. This left me out of the loop of his Ottawa office where all the action was, and I was left to deal mainly with constituency issues. This responsibility involved dealing with constituents' problems and requests, many of which were not solvable, and I was not the person for this job. It was also time-consuming as I had to be in the office five days during the work week, and then on the weekends when Maz was back, I accompanied him on his travels around the constituency.

In 1985, while I was in Maz's Vegreville office, CN Rail invited representatives of the grain industry and other interested parties to visit the new grain terminal at Prince Rupert and then go to Vancouver and take their train through the mountain, travelling in their scenic executive Sanford Fleming guest car. This trip was a great opportunity for me to gain an appreciation of the difficulty in moving grain through the mountains and to view the challenges of operating grain terminals on the West Coast. These types of events gave me opportunities to meet many farm and industry leaders in the Western grain industry.

The trip was organized by Doug Campbell, who was a senior executive with CN Rail. Doug was from Saskatchewan and earlier in his career worked with the Palliser Wheat Growers. He went on to be executive assistant to Agriculture Minister John Wise in the short-lived Joe Clark PC government. This is where we first met in Ottawa.

Working in the Vegreville office offered me access to the minister but did not bring me closer to my objectives of advocating for a deregulated marketing-and-transportation system. The Vegreville responsibilities were time-consuming and not very rewarding. I wanted to become more involved in farm groups that were advocating for change. In early 1986, I left Maz's office and committed my time to the farm. I had become disillusioned and cynical about Ottawa, the bureaucracy, politics, and politicians.

To become involved in grain industry reform, my objective was to become a director of UGG, since it was a farmer-owned grain company with progressive views that advocated for the changes that I wished to see in the grain system.

THE 1980S

The decade of the 1980s was not kind to Alberta and the West. The Trudeau Liberal Government's National Energy Program (NEP) devastated the Alberta economy with spillover effects on neighbouring provinces. The high interest rates of the late '70s and early '80s left many farmers and small businesses deeply in debt. Asset values plummeted and didn't recover until the early 2000s. In 1984, the Mulroney Conservatives were elected, and they abolished the NEP. By that time, world oil prices had plummeted, and the financial pain continued.

Agriculture in the West was also struggling. Grain prices languished due to large world stocks of grain generated by generous domestic subsidies to farmers in the EU and the US. Several dry years led to small crops on the Prairies, which added to the grim financial situation of the farm community.

The railways were now being paid the Crow Gap annually by the government under the terms of the new Western Grain Transportation Act (WGTA) of 1983. Service improved, but efficiencies could not be introduced due to longstanding and continuous regulatory impediments. The legislation did not allow for the closure of low-volume, inefficient branch lines. It forced the railways to charge the same rate for grain shipment whether they spotted one car or a block of fifty. Variable or incentive rates, which would reflect more efficient transportation practices such as multi-car spots could not be introduced.[30] The single car rate was the only rate allowed. Railways were forced to charge the same rate for a single railcar of grain spotted by an old, inefficient elevator on a little used, poorly maintained branch line as they were for multiple railcars spotted beside a new, efficient elevator on a well-travelled, well-maintained mainline.

These regulations carried over from the Crow days and hampered the transformation to an economically streamlined system.[31] Another impediment to operational efficiency was the inability of grain companies to negotiate railcar supply directly with the railroads. Though the grain companies and railroads owned the assets in the industry and carried out the actual activity of shipping grain, the CWB controlled the railcar allocation for wheat, oats, and barley to grain company facilities. Railcars for shipment of non-board crops were now allocated by the Grain Transportation Authority after its establishment in 1979.

The Pools were advocates of the status quo as they had many small elevators scattered all over the Prairies, including many on little used branch lines that were expensive to maintain. Federal statute did not allow the railroads to abandon these branch lines. Many of the elevators on these lines were so small and decrepit that it was not possible to unload a modern multi-axle grain truck in them. Most of these facilities were owned by the Pools at single elevator points where there was no competition.

These mouse-infested relics from the past would be filled with grain and left with storage fees being collected from the CWB. The Pools would then take their CWB railcar allocation and direct the cars to more modern points on railroad mainlines where there was competition from other grain companies. When farmers at these remote elevator points on branch lines would complain that their

---

30 A copy of an article I wrote on the benefits of incentive rates is Appendix #3 to this paper.

31 A well-explained account of the way the WGTA payment to the railways hindered the introduction of system efficiencies can be found in the January 1989 and 1990 issues of *Grainews*. The articles are written by Rick Boyd, assistant VP of CN Rail.

elevator was full, the Pools would simply blame the railroads for not supplying enough railcars. Farmers who were near these points would have to change their delivery point in their CWB permit book and truck their grain further to facilities on the mainlines if they wanted to fill their delivery quota and receive some income.

The Pools campaigned to maintain this old system despite the fact it was costly to the industry. They also supported CWB allocation of grain cars. The CWB/Pool lobby machine was so strong that the government in Ottawa would not challenge them.

New investment in modern elevator facilities was hampered by these regulations as well as the CWB car allocation system. Why would a company build a modern new terminal elevator with a fifty-car spot when you could not control the flow of grain through it or get a lower freight rate for a larger railcar spot for this investment? With the CWB car allocation system based on historical handling, a new facility could not get enough car allocation to fill a large spot. The system was laden with inefficiencies.

By the 1970s, railway branch lines constructed in the 1920s were in total disrepair. On many little used branch lines, the track bed was so poor that railcars could not be fully loaded. The only traffic on these lines was the occasional railcar of grain.[32] From the 1920s until the 1950s, these branch lines were the lifeblood of many rural communities. Livestock was shipped out on rail. All other goods were shipped in and out of these communities by rail. The train was the principal means by which people travelled from community to community. As better roadways were constructed in the 1950s, and the use of automobiles and trucks became more common, use of the railroads for passenger and freight movement diminished.

Loading cattle onto railcars.

32 Appendix #4. Grain rail network in 1985 before branch lines were allowed to be abandoned in 1996.

"Canadian Pacific Railway Kneehill mixed train with locomotive 2521, Alberta." Spring 1946. (CU1231044) by Yeats, Floyd. Courtesy of Glenbow Library and Archives Collection, Libraries and Cultural Resources Digital Collections, University of Calgary.

This era is romanticized by photographers and historians. Idyllic photos of lines of grain elevators are nostalgically remembered as the good old days. It was a bygone era where the Prairie population was largely rural, consisting of scores of bustling Prairie towns and villages with their rows of grain elevators located along railway sidings that criss-crossed the Prairies. This symbolized the bounty of Canadian Prairie wheat farming.

Elevator row at Beaverlodge, Alberta in 1931. Stooks of unthreshed grain are in the foreground. This photo depicts how close to each other the elevators were built with each elevator not having room for more than a two or three railcar spot.

Courtesy of Prairie Towns

The world had moved on. New technologies, large farm equipment, and improved roads and highways resulted in a consolidation of Prairie farms. Without a large rural population to sustain them, many Prairie towns and villages are now gone. The lines of country elevators, too small to accommodate large, modern trucks and large railway car spots are gone, replaced by large grain terminals more fitting for modern times.

THE ALBERTA GRAIN COMMISSION

During this time when we were slowly building the farm, I became involved in various boards and commissions continuing to influence positive change. After leaving Maz's office in 1986, I maintained contact politically, both federally and provincially. I became associated with progressive farm groups, which included the Western Canadian Wheat Growers Association, the Western Barley Growers, and the Alberta Canola Growers Association. These were new farm groups looking for changes to our system of marketing, handling, and transporting grain.

In the early 1970s the Alberta Grain Commission (AGC) was launched by Dr. Hugh Horner, deputy premier and minister of agriculture in the Lougheed government. It was a small, two-man office with an advisory committee of progressive farmers from across the province and a Member of the Legislative Assembly who provided caucus liaison. It was established to advise the minister and caucus on matters related to the grain industry in Alberta. These included agronomic, marketing, and transportation issues, among others. The first chairman of the AGC was John Channon, a former federal civil servant. Neither Doc Horner nor John Channon were supportive of the CWB monopoly or the regulated grain-marketing-and-transportation bureaucracy.

In the summer of 1986, I began to work in the AGC office. Later in 1986, I was appointed as a farmer member by Agriculture

Minister Peter Elzinga. This was a great opportunity to put forth my views and use the experience I had acquired on the farm, working in Ottawa, and with the Grain Transportation Authority. The Alberta Grain Commission was aggressive in its policy recommendations, and the Lougheed government in Alberta was open to accepting them. Many issues discussed at our table worked their way up to become government policy. I had the opportunity to meet and work with other AGC farmer members, too many to mention. Jack Gorr, Glen Goertzen, and Ken Beswick were among those I enjoyed working with. We shared common views on what a deregulated marketing-and-transportation system should look like.

From 1989 to 1993, Ernie Isley was Alberta's minister of agriculture. He was not supportive of Winnipeg's grain-related bureaucracy and would make trips to Winnipeg to lobby the industry for constructive change. Accompanying him would be senior officials from Alberta Agriculture and some AGC farmer members. A trip with Minister Isley was always good for a few games of cards and ample refreshment.

In the late 1980s, Ken Beswick became a member of the AGC, and then left his farm in southern Alberta to become chairman of the AGC. In 1992, Ken was appointed one of the five commissioners of the CWB by the Mulroney government.

I served on the AGC from 1986 until 2002.

Looking back at my time on the AGC, and the achievements of the group, Glen Goertzen reminded me of the following quote:

Never doubt that a small group of thoughtful, committed citizens can change the world. Indeed, it is the only thing that ever has.

–Margaret Mead, scientist

ALBERTA GRAIN COMMISSION
1995 Members

Top: from left to right

Gil Balderston (Producer Representative - Sexsmith), **Ken Motiuk** (Producer Representative - Mundare), **Ray Bassett** (ADM, Planning & Development - AAFRD), **Glen Goertzen** (Producer Representative - Stettler), **Dan Cutforth** (Producer Representative - Barons)

Bottom: from left to right

Pat Durnin (Producer Representative - Kathym), **Murray McLelland** (Cereal Crop Specialist - AAFRD), **Ken Moholitny** (Chairman - Alberta Grain Commission), **David Walker** (Executive Secretary - Alberta Grain Commission), **Eugene Dextrase** (Producer Representative - High Level)

FARM DEBT REVIEW BOARD

Another federal body I became involved with in the 1980s was the Farm Debt Review Board (FDRB).

The 1970s were a boom time for grain and oilseed farmers. The decade started out with burdensome world supplies and low prices. Then along came the Soviets in 1972, and with great stealth and secrecy, with the market unaware, managed to buy up much of the free stock of grain from various merchants in North America. When the magnitude of these purchases became known, prices skyrocketed. Wheat prices tripled in a little over year in 1972–1973. Land prices, assisted by high inflation rates, also boomed. Farmland selling for $100/acre in 1971 reached prices of $1,000/acre in 1981. Inflation exceeded 10% in many years. To combat inflation, central banks raised interest rates to slow the economy, and in the early 1980s interest rates were up to 20%.

Farmers worldwide responded to these higher prices, and by the mid-1980s the situation changed substantially. Crop production increased, stocks went up, and prices went down. The European Union and the US introduced various price-support programs to subsidize their producers, so farmers kept growing big crops. Canadian farmers did not have these types of guaranteed price programs and were faced with grain prices that were much lower than in the 1970s. A perfect storm of negative events developed. Farmers who paid high prices for land in the late 1970s faced large mortgage payments they could not meet. Adding to this were the very high interest rates that compounded rapidly when a payment was missed. Land prices started to fall as there were no buyers. Additionally, the Federal National Energy Program and the lower oil prices in the mid-1980s decimated the Western provinces' economies, and all real estate values fell substantially. Farmland that sold for $1,000/acre in 1981 fell as low as $500/acre by 1986.

Many farmers (and other real estate owners) soon found their large mortgages from the boom times were "underwater," that is, more was owed on the land than it was worth on the market. A financial equity crisis was building as farmers were unable to make payments on these mortgages and financial institutions found themselves with many nonperforming and impaired loans.

In response to the deteriorating financial position of many farmers, in the mid-1980s the Mulroney government set up the Farm Debt Review Board (FDRB). This was a mediation service for farmers and their creditors. Each province had its own FDRB. A small number of well-established farmers and businesspeople formed the board of directors, and several contract support staff with financial management experience were hired. When a farmer faced remedial financial action from a financial institution, an application could be made to the FDRB. Once an application was received, the FDRB froze all assets and liabilities of the farmer for 120 days. A contractor was assigned to the file and visited the farmer and creditors to prepare a report.

After the report on the farm situation was prepared, a mediation panel meeting was set up between the farm owners and the creditors. This panel meeting was chaired by a senior FDRB member, the contractor who prepared the report, as well as two other contractors to assist in exploring mediated options for the parties involved. The entire process was voluntary from a solution perspective, and the objective was to keep farmers out of expensive financial litigation.

I worked as a contractor with the FDRB from 1985 to the mid-1990s as I was becoming more involved with other farm organizations. The FDRB was a tremendous learning experience, once again meeting and working with many knowledgeable people in the agriculture industry. The workload fit well with the farm as there was little FDRB activity in the busy seasons. I also learned a tremendous amount about mediation, financing, and financial management. Witnessing the poor decisions made by financially illiterate farmers and the consequences of them was a sobering and practical exercise. My entire experience at the FDRB later served me well as we were building our farm.

During this entire period, I was becoming more involved in organizations and associations where my goal was to achieve a market-oriented system of marketing, handling, and transporting grain. I was also trying to gain exposure to run for a UGG director position when an opening arose. I became active on the Mundare UGG local board and attended UGG annual meetings as the delegate from Mundare.

SENIOR GRAIN TRANSPORTATION COMMITTEE (SGTC)

Through the 1980s, the grain industry had an entity called the Senior Grain Transportation Committee. The SGTC was comprised of senior members from the system and included several elected farmers. The committee was earlier established as a forum for system participants to gather regularly and address issues and problems in the system. Having senior executive members attending provided more assurance that courses of action agreed upon at a meeting would be implemented.

The farmers on the committee were elected by the farm community. Most of these elected farmers came from the Wheat Pool community, as the Pools campaigned to elect farmers supportive of Pool policy. The SGTC became strongly divided along political lines. The Pools, CWB, and most elected farmers supported the status quo, a regulated approach to system management. Other members of the SGTC, including the railroads, private grain companies, UGG, and some elected farmers, supported a market-based approach.

In 1988, Glen Goertzen, a farmer from the Stettler, Alberta area and I were elected to sit on the committee. Glen and I became allies and good friends as our views on the future of the system meshed well. We were persistent advocates for change and challenged the old ways of thinking that dominated the committee. We could never get enough support to change policies, but we were able to broaden discussion and advocate for new solutions to the old problems.

It was a bit intimidating at first, sitting around the table with senior executives representing all facets of the grain-handling-and-transportation system. However, Glen and I did our research and contributed to the discussion. Many from around the table welcomed our blunt questions and suggestions for change. Others were afraid to say anything for risk of retribution from the CWB. Our participation almost always resulted in a critical response from either of two men around the table—Larry Kristjansen, Commissioner of the CWB, or Ted Turner, President of Saskatchewan Wheat Pool (SWP). The elected farmers that challenged them were treated arrogantly and condescendingly.

Their attitude towards the independent thinking elected farmers reflected the intensity of their drive to preserve the status quo. Kristjansen, a dour, portentous sort, would stare at us over the glasses perched on the end of his nose. He would proceed to lecture us like rural peasants reaching beyond our station in life, informing us that we did not know enough to question things, and did not appreciate what the CWB and the Pools did for farmers. This was a good example of how men of modest self-accomplishment talk themselves into positions of power and then use intimidation to dominate others.

The real problem was that even though senior members from the industry had the power to change things, the committee would only entertain discussion on operational issues rather than broad policy changes. Glen and I were not afraid to challenge this, and there were many spirited discussions. Since bad policy was the root cause of problems in the system, the committee was ineffective in accomplishing change.

I sat on the Senior Grain Transportation Committee as an elected farmer until I was elected to the board of UGG in 1990. I then remained as UGG's representative until the committee was disbanded in 1995 when the WGTA was replaced with the Canada Transportation Act (CTA).

During this time in the late 1980s I also became involved with the Western Canadian Wheat Growers Association (WCWGA).[33] This was a voluntary group of Saskatchewan-based farmers who strongly believed that the grain marketing and transportation system should be driven by market forces, not regulatory control. The WCWGA was a great forum to meet many progressive and forward-thinking farmers and promote my views on changes necessary in the system. I rose to the rank of Alberta Vice-President and left the association when I became a UGG director. It was here that I met Hubert Esquirol, a tireless advocate for change. We became friends and allies as we represented the WCWGA and promoted change at any venue we could get invited to.

33 Originally Palliser Wheat Growers Association.

Western Canadian Wheat Growers Board of Directors - 1991

PRO-FARM

The Official Publication of the Western Canadian Wheat Growers Association

Vol. 7 No. 1 February 1991

Diversifying in the '90s
Convention Highlights
GRIP and the Marketplace

1991 Western Canadian Wheat Growers Board of Directors

Back row, left to right: Peter Edgar, Dan Quark, Larry Maquire, Len Rutledge, David Fulton, Warren Jolly.
Front row, left to right: Ken Motiuk, Hubert Esquirol, Jack Gorr, Harvey McEwan (President), Greg Downie, David Rose.

MY GAWD . . . WE HAVE LOST OATS[34]

In 1984 the Mulroney Conservative government came into power in Ottawa, and progressive farmers were hoping to see some change. The Mulroney government was at times frustrating to those of us who strongly supported his government and wished to see positive change in the system. Western farm opinion on marketing and paying out the Crow Benefit was polarized. Many farmers who supported the Conservatives also supported the CWB and the Pools and did not wish to see change in the system. Little change occurred for most of the Mulroney era. The Crow Benefit continued to be paid to the railroads, regulations hampered the introduction of efficiency measures into the system, and the CWB maintained control over marketing Western grain and railcar allocation.[35]

On the Prairies, there were also many dry years in the 1980s and along with low world prices for grain, the government focused on farm financial assistance programs. The Mulroney government was re-elected in 1988 with a platform promoting a free trade agreement with the US. Charlie Mayer was appointed minister responsible

---

34　In the March 6, 1989 issue of *Grainews*, Lyle Walker, an Alberta farmer wrote a tongue-in-cheek article mocking the outcry of the Left over taking oats off the CWB. It was titled "My Gawd . . . We Have Lost Oats!".

35　The benefits of a pay-the-producer resolution of the WGTA vs pay-the-railroads is well researched and spelled out in a 1990 publication from Alberta Agriculture, "*Paying the Producer—Good Today, Better Tomorrow*".

for the Canadian Wheat Board. Mayer was not supportive of CWB monopoly marketing and felt that farmers should have a choice. Meanwhile, the number of Prairie farmers seeking change was growing.

In the mid-1980s representatives of General Mills started to visit Alberta farmers, as they were wondering why they were not able to source oats from Western Canada. General Mills was one of the largest purchasers of oats from Western Canada, and makers of well-known Cheerios. We had started growing some oats on our farm and found we could grow high-quality oats that commanded a high price in American markets. It was also a good crop to fit into our crop rotation.

The CWB was not interested in a small-market crop like oats. While customers were looking for sources of Canadian oats, little or no oats were being exported by the CWB. The following graph shows oat exports from 1949 to 2022. (page140) The dismal performance of the CWB in oat marketing in the 1980s is evident in contrast to the success of the open market in more recent years.

Executives from General Mills in Minneapolis visited the Motiuk farm in 1989 to check out the oat crop. They hoped the introduction of private marketing of oats would help them get sufficient supply from Canada since they were having difficulty buying oats from the CWB.

Included in the photo with General Mills executives are Alberta Associate Agriculture Minister Shirley Cripps; Alberta oat growers Lawrence Kapicki, Nickolas Jonk, Peter Kirylchuk, and Ken Motiuk, with daughter Erin Motiuk in the foreground.

At the 1989 WCWGA annual meeting in Calgary, CWB Minister Charlie Mayer informed me that the government was planning to remove oats from the jurisdiction of the CWB. He knew there would be an outcry from all left-wing farm groups and requested some help in organizing public support from farmers and farm groups who agreed with the decision. Several progressive oat growers banded together and formed the Alberta Oat Growers

Association (AOGA). The group came out loud and clear in support of free marketing of oats to offset the anticipated outrage from the left-wing farm groups. The first president of the AOGA was Peter Kirylchuk from Lac La Biche, AB, who did a great job in getting the fledgling organization established and operating.

Sure enough, as soon as Mayer made the announcement he would remove oats from CWB jurisdiction, the Pools and NFU denounced the move. They used hyperbole, predicting this would be the beginning of the erosion of the CWB to the economic demise of Western Canadian agriculture. The AOGA, UGG, and other market-oriented farm groups supported the move. Mayer and the Mulroney government stuck to their resolve. Finally, we had a government willing to challenge CWB power over Prairie grain marketing.

The CWB's performance on oat exports had been dismal. From December 1980 to December 1981, the CWB exported no oats at all. In 1985, they exported 10,000 tonnes, which is clearly shown in the graph below. Meanwhile General Mills couldn't get sufficient supplies of oats from Canada for their milling program. In the past decade, the grain trade has exported an average of over 1.2 million tonnes of oats annually, with over 2.0 million tonnes exported in some years.

Annual Oats exports from Western Canada in millions of tonnes

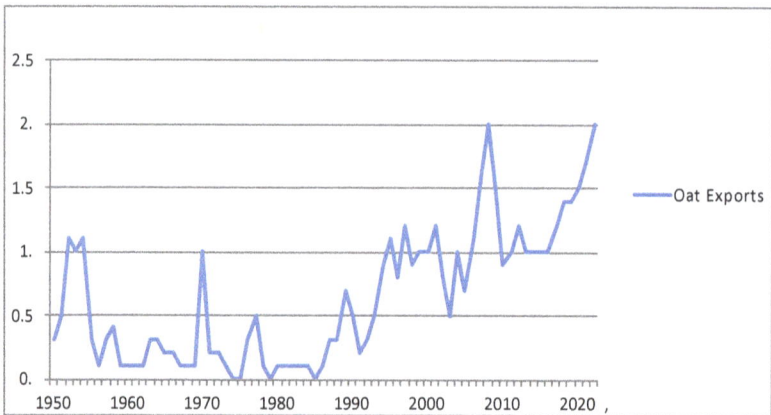

Source: Canadian Grain Commission

In 1949, oat marketing was placed under CWB jurisdiction by Order in Council (OC). The legislation of the CWB Act was not changed at the time. Since oats went under the CWB by OC, it was possible to take it out in a similar manner. This was much easier than having to go before Parliament and change the Act. A simple Order in Council rescinded the 1949 Order and accomplished the task of putting oat marketing back on the open market.

Once the private trade took over, oat exports increased substantially. Today oats are a small but profitable crop on the Prairies.

≈

Throughout the Mulroney administration, the 1983 Western Grain Transportation Act payment to the railways of the Crow Gap was a contentious debate in Western farm circles. It was polarizing and controversial. The Pools wanted the payment to continue to the railroads while the market-oriented farm groups wanted it paid directly to farmers.

There was some discussion in back rooms about a third option, which was to simply pay out the feds' annual commitment through a one-time, upfront capital payout to all Western farmers. A total sum of $6.7 billion was informally and unofficially spoken of by select government members. This would serve as a one-time buyout of the federal commitment to subsidizing grain freight rates.

When the Pools learned of this, they strongly opposed it because they did not wish to see the regulated rate structure change. Once again, as when the first version of the WGTA was proposed in 1983 by Gilson who advocated a payment of the subsidy directly to farmers rather than the railroads, the Pools attempted to stop this initiative. They went to Quebec farm groups and leaked the discussion that the West could potentially get all this money and Quebec would not get any. This unholy alliance of Quebec farm groups and the Pools killed the idea of a one-time payout to farmers in the order of $6.7 billion since the Mulroney government chose not to have this political battle.

Late in the Mulroney term, CWB Minister Mayer attempted to act on the CWB monopoly over barley marketing. Time proved that his earlier action on oat marketing was a commercial success. Unable to make any progress advocating for removing barley completely from the CWB, progressive farm groups started advocating for a continental barley market, where barley could be sold to the US without going through the CWB. The idea was that since so much of our barley went to the US, and a free trade agreement was now in place with the US, farmers should be allowed to ship barley directly there. It was a half-measure at best, but it made economic sense. Mayer issued an Order in Council (OC) that allowed Western farmers to ship barley directly to the US without going through the CWB. CWB jurisdiction over barley exports to other countries remained.

Predictably, the Pools and the National Farmers Union opposed this. The Saskatchewan Wheat Pool challenged the action in court and won. The distinction was that with oats the Order in Council worked because it was a complete removal of oats from CWB jurisdiction. The judge ruled that the Order in Council on barley was not legal because the government could not do a partial removal—it was all or nothing. It was not legally possible to allow barley exports into the US while leaving the CWB in charge of barley exports elsewhere. So, barley marketing went back to where it was, all under the CWB. It was a noble effort by Mayer, but the SWP's legal challenge thwarted this change. It demonstrated how the Pools and the CWB acted in concert to preserve their dominance in the Prairie grain sector.

Under existing CWB regulation, if a Western farmer wanted to sell a load of barley to the US, he would first have to sell it to the CWB at the initial price (usually quite low). Then he would have to buy the same barley back from the CWB at a much higher price, the difference going into the barley pool account. There was no way to ensure that the buyback price was fair since it was all done behind the closed doors of the CWB. In fact, when a farmer arranged his own sale and by the time he went through the CWB machinations, the buyback price was so high that it left no profit for the farmer willing to find his own market. This sequence of CWB-regulated events discouraged farmers from shipping directly to markets across the border on their own. Farmers who violated this process were subjected to legal action, fines, and incarceration.

In the early 1990s, Eric Malling from CTV's news program W5 was working on a story on Don Mazankowski. I met Malling in Maz's office in Vegreville. Malling was from Saskatchewan and was somewhat familiar with the Prairie grain industry. I explained to him the process the CWB required of a farmer before he could sell his own barley directly to the US. Malling was astounded with the whole procedure. He suggested doing a segment on W5 exposing this to Canadians. I introduced Malling to Buck Spencer, a former UGG director and farmer from southern Alberta who had been trying to sell his barley directly to the US and was having difficulty with the CWB. Spencer and Malling got together and W5 aired a program on Spencer going through the regulations the CWB required of him to sell his own barley into the US.

FINALLY . . . UGG

When I left Maz's office in 1986, my objective was to get on the board of directors of United Grain Growers. The way UGG elected their farmer board members is important to understand since this was the basis of their governance structure. UGG was governed by twelve elected directors who were all farmers. The decisions these twelve farmers made guided the operations and future investments of the company. The Pools were similarly governed, but they attracted farmer-directors with a socialist mindset that called for government intervention and CWB control. UGG tended to elect farmer-directors who believed in market-based solutions with more freedom for farmers to make their own decisions. This was the path former presidents Tom Crerar and Mac Runciman had set UGG upon, and that was why I was drawn to UGG.

At every country elevator location, UGG would form a local board. This was a group of farmer-customers who would advise the company of issues at their individual location, as well as provide recommendations on broader farm policies to the board of directors in Winnipeg. These locals were assisted by a team of UGG staff called field services representatives, who were the conduits between the local boards and the Winnipeg main office. Each year a farmer representative was elected from each local to attend the UGG annual meeting, where the company reported on its financial affairs and discussed policy issues along with company involvement in other matters in the industry.

At the annual meeting, the local delegates voted for farmer-directors—four directors annually on a rotating basis, all for three-year terms. Arguably, it can be said that voting for UGG directors was the most important thing the local delegate did. I had been the local delegate from Mundare several times prior to being elected a farmer-director on the board of UGG.

In 1990 the opportunity arose for me to run for director. The annual meeting was always held in November, and that year there were three vacancies to fill on the twelve-member farmer board. Chairman Lorne Hehn from Saskatchewan left, as he had been appointed chief commissioner of the CWB. Meryl Layden and Buck Spencer, both from Alberta, had retired. That August, at a GATT (General Agreement on Tariffs and Trade) meeting in Ottawa where I was representing WCWGA and Ted Allen was representing UGG, Ted encouraged me to run for UGG director at the upcoming annual meeting. I thanked him for his encouragement and told him my plan was to run.

Many candidates put their names forward in the November 1990 election, and when it was over, Terry Youzwa was the new director from Saskatchewan and Bernie MacKay and I were the new directors from Alberta. I was elected on the first ballot in the 50%-plus-one voting system that was used for UGG director elections. Other directors were elected in the later ballots. It was an exciting time as there were so many new faces on the board of UGG. Ted Allen, always an advocate of market-based solutions rather than the grain-regulatory bureaucracy, became the new chairman of the board. Brian Hayward, a young executive in the company, became the new CEO. At the time, the company was not enjoying great profitability. It had a smaller and aging line of elevators in the country and was soon facing paying out some large patronage holdings.

This was the world I was cast into in the fall of 1990. It was the real world with responsibility for real dollars and cents, not just some advisory or policy body. A director of UGG carried a fiduciary duty where the director needed to act in a manner that would benefit someone else financially (i.e., the shareholders of UGG). The entire industry was facing a pent-up need for reinvestment and consolidation that was being hindered by the regulatory environment. Once the regulatory environment changed, it would totally transform the way the system operated. In a decade and a half, the outcome would be something nobody had imagined or envisioned. But we did not know that yet!

UGG had several business units doing various things at the time, and they did not all operate cohesively. Frankly, the company had not been run well for the past while under the leadership of President and CEO Lorne Hehn and COO Gerry Moore. It was difficult for grain companies to be profitable since the system was so regulated and it did not allow for profitable opportunities to arise. Often in the search for opportunities to make money, grain companies ended up in ventures where they had no expertise and were not related to their core business.

≈

In 1991 there were six major line elevator grain companies and two smaller ones in Western Canada. The largest companies were the Pools with about 60–65% of the business in each province. UGG had about 13% of the total Prairie market with Pioneer Grain (Richardson) having about 11% and Cargill 10%. Manitoba Pool Elevators (MPE) was headquartered in Winnipeg, Saskatchewan Wheat Pool (SWP) in Regina, and Alberta Wheat Pool (AWP) in Calgary. United Grain Growers (UGG), Pioneer Grain, and Cargill were all based in Winnipeg. The Pools and UGG were farmer-owned companies with their boards of directors elected by farmer-owners/-patrons. Pioneer was privately owned by the Richardson family and Cargill was a large, private American company owned by the Cargill and MacMillan families. The two smaller companies were Parrish & Heimbecker and Patterson Grain, both family-owned companies based in Winnipeg with a combined market share of less than 5% of the business.

At this time, most grain elevators on the Prairies were old, small, and run-down wooden facilities. They were scattered all around the countryside on both main lines and little used branch lines, each with small car spots. It was a system designed for the 1920s, when grain deliveries were being made by horse and wagon.

Because of all the regulations, there had been little investment in new facilities over the past number of years. The CWB controlled the throughput of the country elevators with their delivery quotas and railcar allocations. The payment of the Crow Benefit to the railways resulted in them being more eager to pull grain, but the costs in the system were rising and the Crow Benefit was fixed. This resulted in farmers beginning to pay more for freight. The system needed to be modernized to include large terminal facilities with large car spots on main lines for efficient rail movement.

The small, old grain elevators, owned mainly by the Pools, were inefficient. Most of these elevators had been built quite close together in a row on rail sidings. Because these elevators were built so close to each other, railcar spots were usually under three cars. Many of the older elevators could not handle the larger trucks farmers were now using. Weigh scales were too small. The unloading pits were small and the elevator legs slow. The ceilings over the unload pit were too low for the longer boxes on tandem-axle trucks to be able to be raised high enough to empty the loads. The driveways were too tight and narrow to allow semi-trailers loaded with grain to pass through. The CWB car allocation formula preserved this old system.

The CWB distributed railcars between grain companies using something they called the block-shipping system. The Prairies were divided into blocks of track that would roughly make up a train run. There would be many elevators, both large and small, in a block. When a train of one hundred cars was distributed, the CWB would allocate the cars to the various companies in the block based on their historical handling percentage. Each company would then choose which of their elevators in the block would get the cars.

The Pools with a history of about 65% of the grain handle would get 65% of the car allocation. If UGG historically handled 13% of the grain in that block, they would get 13% of the cars allocated, and so forth for the other grain companies. The problem with this system was that no company could increase their market share in any way, since the formula was self-perpetuating. This locked in the market share for every company with the Pools dominating grain collection on the Prairies. This arrangement was favourable for the Pools. They would fill all their small, old elevators with grain and then direct the car allocation to the point where they had a larger facility and more competition from others. This made it almost impossible to change market share for the companies.

The other deterrent to a more efficient allocation of grain cars was the fact that the railways were forced by the WTGA to charge the same on a per-tonne basis to ship from a new, efficient facility on a main line with a twenty-five-car spot, as they charged on a tiny, dilapidated, old facility on a little used, overgrown branch line with a two-car spot. No grain company was rewarded for an economic investment in a large new facility with a large railcar spot.[36]

> "The hypothesis put forward is that co-operatives in Western Canada had, prior to the 1990s, been operating in an economic environment that was highly regulated and thus, relatively stable. This environment changed in the 1990s, and because of this change, management was given additional power and influence. Management was also overconfident – the co-ops that they managed had performed relatively well in the 1970s and the 1980s, and they believed that this strong performance was due to their management skills, rather than being a consequence of the protected regulatory and economic environment in which they operated."

---

36   Fulton, Murray and Larson, Kathy. *Overconfidence and Hubris: The Demise of Agricultural Co-operatives in Western Canada.* AgEcon Search. 2009. Page 180.

The CWB car allocation system did not allow companies other than the Pools to utilize a twenty-five-car spot unless they had 25% of the grain handle in a block. Since the companies all had significantly lower market share, they could not get a large car allocation from the CWB. Why would any one of them build a large, new, rail-efficient facility if you could never get the cars for it from the CWB? The Pools were the only companies with a car assignment from the CWB large enough to be able to capitalize on the efficiencies of a large car spot. They were also philosophically opposed to building large Prairie terminals. Instead, they were capitalizing on getting everything they could out of their large number of small elevators, benefiting from the fixed freight rate of the WGTA, the car allocation system, and CWB storage payments.

After the WGTA was passed in 1983, the railways began to invest in improving their grain-hauling infrastructure, grain cars, and locomotives. The grain elevator system was quite another thing. After decades of little investment, the line elevator system in the country was in dire need of upgrading and renewal. The farmer-owned entities, the P

ools and UGG, with more than 75% of the volume of grain delivered, were badly undercapitalized. Their equity base was farmer-investors and patronage dividends that had to be paid out. They could not afford to do this and still invest in new facilities. They had no way to raise outside share capital. Debt financing, if they could get it, would be a disaster to their balance sheets. The private family firms, including Cargill, were not willing to invest more money in facilities they could not control. The same was true for new investment from foreign sources.

A FULL PLATE

Meanwhile, between the farm and everything else I was involved in, I now had way too many things going on. When I was elected to the board of UGG, I was on the Alberta Grain Commission, serving as Alberta VP of the Wheat Growers, and doing contract work for the Farm Debt Review Board. We were also growing the farm. Almost weekly, I travelled all over the West lobbying and going to meetings. We had four young daughters at home and Wendy was juggling farm work and a casual nursing job while doing most of the parenting. I left the Wheat Growers and scaled back my FDRB work.

We did lots of juggling and thank goodness for the energy of youth. In 1992, Wendy and I were proud to win the Outstanding Young Farmer award for Alberta. We were honoured to be acknowledged for the progress we were making on the farm and be recognized by the larger rural community. We felt good about our accomplishment.[37]

---

37 Appendix #5

Our farm was continuing to grow. Our seeded acres were by then at 2,100 and my aging father was doing less on the farm. Wendy had to step in and help in the busy seeding and harvest seasons. Our long-term objective was to grow the farm so it would become large enough to provide us with a good standard of living without having to work off the farm. Every time we had a chance to get more land to farm, we took it, either through purchase or lease.

We also introduced new and innovative changes on the farm. We started seeding dry field peas. Peas were another crop where there were no quotas, and we could manage sales and cash flow. We adopted the practice of no-till planting of crops as new equipment technology had developed planter units that did not require pre-cultivation of the soil. This was a time-saving move that eliminated the cost of a great deal of cultivation equipment and resulted in reduced costs of fuel and labour. Glyphosate herbicides managed weed control. All crop residue was returned to the soil to improve organic matter and fertility.

We started straight-cut harvesting our cereal grains in the '80s, eliminating the need to swath these crops in an extra pass across the field. In the early '90s, we started straight-cutting canola, which eliminated the need for a costly swather on our farm and the extra time needed to swath. Early adoption of these new cost-saving practices gave our farm a financial and competitive edge while neighbouring farmers maintained more costly traditional practices.

We were getting good crops and needed to store more wheat. While we waited for CWB quotas to open, we were annually building new storage bins. We also needed additional farm equipment storage sheds. All this required a great deal of cash. Grain prices were poor in these times and cash flow was important.

Canola acres were increasing on the Prairies and canola became known as a cash crop. This meant you could produce the crop and sell it for immediate cash at harvest or whenever you wanted. There were now no quotas on canola delivery, so you could plan your cash flow after harvest. You could contract canola with a grain company and lock in a price for a future delivery date. This was totally in contrast to the CWB pooling system, where it would take up to sixteen months after harvest to receive all your money.

This was the progress we had made on our farm when I was elected to the UGG board of directors in November of 1990.

THE UGG BOARD

Joining the UGG board was like joining a big family. There was an informal order of seniority around the table, based on length of service. I started at the far end of the table like the other rookies. Annually, UGG would have a summer meeting at a resort type of facility, where all the directors would bring their families for several days. Each Christmas, the president held a Christmas banquet for directors and their wives. Spouses always attended the annual meeting. During the summer board meeting, many of us would go golfing together and good friendships evolved.[38] It was much more cordial and social than the atmosphere of a usual board of directors.

---

38 Appendix #6

United Grain Growers Board of Directors - 1991

Left to right standing: Terry Youzwa, Sam Sich, Bernie MacKay, Joe Omichinski, Dalton Hockley, Ken Motiuk, Richard Phillips.
Left to right seated: Don Dobson, Bryan Perkins, Ted Allen, Roy Piper, Roy Cusitar.

As a new director, I faced a steep learning curve. I came to appreciate and respect Bryan Perkins, a farmer from Wainwright, Alberta, who was Senior VP and a mentor to me on the board. He was astute and well-spoken, had good business acumen, and was an excellent judge of character. I learned a great deal from Bryan, and we still stay in touch.[39]

---

39 After leaving UGG, Bryan went back to Wainwright and, along with family and business associates, built a large grain farm as well as a large, integrated pork-production business. He remains at the helm of this today.

Left to right standing: Ken Motiuk, Wendy Motiuk, Bryan Perkins, Sharon Perkins, Roy Piper, Elaine Piper.

After the 1990 election, the new board at UGG focused on streamlining some of the business units, putting new management personnel into place, and identifying some common unifying thread that would focus on future growth and profitability. One of the larger issues we had to deal with was the capitalization of the company. Capitalization of co-ops and member-owned companies such as UGG is always difficult, as most of the investment equity is made up of either patronage dividends owed to the owners/customers or investment shares owned by the same. It was problematic in that these sources of capital were promised to be paid out when a farmer retired. Sources for replacement equity had to be found. This entire issue of equity structure would come to fatally plague the grain co-op movement in Western Canada over the next decade.

UGG was the first of the farmer-owned grain companies to address the issue of equity erosion due to patronage dividends (pat divs) and farmer investment shares that had to be paid out. The objective was to raise capital through outside investors, essentially becoming a publicly traded company. A problem lay in the way UGG was structured. UGG operated under a special Act of Parliament that defined its charter and the way it could operate. To go public, the UGG Act would have to be amended. This change would have to go through Parliament. This was a minor issue in the bureaucratic mire of Ottawa, and it would be difficult to get such an issue on the government agenda.

Fortunately, the Mulroney government was in power, with Don Mazankowski as deputy prime minister. He was familiar with UGG and understood the issue and the industry. Having worked for him previously, I spoke with him about this. He asked to be provided with all the relevant information and he would see to having the required amendments passed. In 1992, in the final year of the Mulroney term, thanks to Maz and his staff, the new UGG Act was proclaimed, and UGG became a public company.

This was a historical event as it was one of the first times a co-op ever went public. This enabled the company to convert over $20 million of farmer-customers' patronage dividend debt into equity. In addition, 2 million common shares were sold to the public at $8.00/share, and after paying out the farmers who wished to sell their shares, another $8.8 million went to UGG cash reserves to further strengthen the balance sheet.

The amendment to the Act allowed for a change in the governance structure of the company. Previously the board of directors consisted of twelve elected farmers. The amended Act allowed for the addition of three non-farmers to the board. It was recognized that individuals who had broader experience in corporate governance would be an asset to the board. In 1993, Jon Grant from Quaker Oats; Arthur

Mauro, former CEO of Investors Group; and Bill Woodward, director of Reed Stenhouse, joined our board.

We were positioning ourselves to grow the company.

United Grain Growers Board of Directors – 1997

Left to right standing: Maurice Lemay, Bernie MacKay, Jon Grant, Ernie Sirski, Arthur Mauro, Wayne Drul, Spence Sutter, Ken Motiuk, Bill Woodward.
Left to right seated: Terry Youzwa, Bryan Perkins, Ted Allen, Roy Piper, Henry Penner.

THE WGTA IS GONE AND THE SYSTEM
STARTS TO TRANSFORM

The spark that ignited the rapid change that was to come in system rationalization occurred under the minority government of Prime Minister Paul Martin and Minister of Transport Ralph Goodale in 1995. As a budgetary measure to control government spending, Martin's Liberal government decided to repeal the WGTA and stop paying the Crow Benefit to the railways. The Canada Transportation Act was passed in January 1996 to replace the WGTA. This brought an end to direct transportation subsidies but continued the government's role in setting maximum freight rates. Railway branch lines were no longer protected from abandonment. Changes in the legislation now allowed for more substantial variable rates for large railway car spots. Rapid consolidation and asset transfers of country elevators commenced. Larger car spots and regional consolidation became the industry drivers in the race to adjust to new conditions. This was truly a positive, pivotal event in the deregulation of the Prairie handling-and-transportation system.

Farmers were about to see a large increase in freight rates, which would no longer be subsidized. The government paid out the benefit by compensating farmers with a one-time capital payout of $1.6 billion dollars. There were no consultations, no studies, no debate, and no negotiation. It was announced as a fait accompli by the Liberal government in their budget. A short-term upward tick in grain prices for a couple of years after the change absorbed much of the sticker shock of the increased freight rate to farmers and helped mask the freight cost increases to the political benefit of the Liberal government. The Paul Martin government, as with most Liberal governments, had few elected members in Western Canada.

In the early 1990s, as they did in 1983 with the Western Grain Transportation Act's method-of-payment issue, the Prairie Pools again had aligned with Quebec farm groups to prevent changes to the rail-regulation system. They killed the idea of a potential buyout of the Crow Benefit by the Mulroney government to Western farmers of near $6.7 billion. In 1995, the Pools stood helplessly by as a much smaller cash settlement of $1.6 billion was imposed upon Western farmers by the Martin Liberal government.

After this, just like pulling a bottom orange out of a pyramid stack in a grocery store, the high-cost, regulated system started to collapse. Market forces started to take effect, and this unleashed a flurry of activity as all the grain companies scrambled to capitalize on the new environment. A commercial rate structure exposed the high-cost inefficiencies in the system. The dramatic transformation that occurred over the next decade and a half resulted in a system that by 2008 was unrecognizable from that of the past. For the most part, the change was positive for farmers and all members of the industry.

The new environment allowed for meaningful discounts to be introduced for multi-car spots, and these variable freight rates began shaping a new and more efficient rail system. Most current facilities did not have a large enough car spot to capitalize on these savings. Grain companies started swapping elevators with each other at individual points to be able to achieve larger car spots. At this time, most towns still had an elevator row, a line of elevators owned by different companies, each with less than a three-car spot. (The 1979 aerial photo of Mundare on (page 78) illustrates this well). If swaps for mutual benefit could be made, one company could take over the line of facilities in one location in exchange for another. This made for more efficient operating units as an entire line of cars could now be spotted for one company at one location alongside several facilities.

However, the core issue of CWB control over railcar allocation along with control of throughput at the port terminals remained unchanged. Grain companies and railways still could not control the flow of CWB grain through the facilities they owned. This key hindrance to system efficiency would not be remedied until the CWB lost its monopoly in 2012.

Grain companies began a surge in construction of large, new country terminal facilities, replacing the old facilities on elevator row. This required a great deal of capital, and UGG, having raised funds in the equity market, had a strong balance sheet and some cash to start building new terminal facilities. The Pools were still hampered by their weak balance sheets and the large patronage dividends they soon had to pay out.

WINNIPEG COMMODITY EXCHANGE

From 1993 to 1996, I served as a public governor on the Winnipeg Commodity Exchange (WCE), a futures market where commodities were traded. The governance structure was a board made up of Exchange members, including representatives of grain companies and exporters, canola processors, and smaller independent floor traders (locals) who provided liquidity in the daily trading activity. The board also included a position for a non-aligned member who was meant to represent the public interest. This position was open in 1993 and the board, through President Fred Siemens, asked if I would serve. I had come to know Siemens in the Winnipeg grain circles. I thought it would be an interesting experience, so I accepted.

The WCE had been in slow decline for several years. Efforts to introduce more crops to be traded were not successful due to a low volume of trade that lacked sufficient liquidity in the trading contracts. The WCE was essentially a one-trick pony, with canola being the only crop traded in enough volume to be viable. The CWB didn't use the Exchange, as milling wheat was not traded, only feed-grade wheat, and the Prairie Pools would minimize trading volumes by offsetting overnight purchases with sales, thereby eliminating the trading volume from the pit.

The Winnipeg Grain Exchange, as it was originally called, was established in 1887 as an open forum where buyers and sellers of wheat could trade in a transparent manner. It was successful and by the 1920s had become the pre-eminent price-discovery mechanism for spring wheat in the world. Its activities were severely curtailed when the CWB took over all wheat trading in the West during World War II. The Grain Exchange added rapeseed futures in 1963. In 1972, branching out with the addition of gold futures, the Exchange became the Winnipeg Commodity Exchange (WCE). When it adopted an electronic trading platform in 2004, it became the first fully electronic exchange in North America and the old open outcry trading was shut down.

Before futures markets were traded electronically, trading occurred in a very large, eight-sided, raised ring. The octagonal ring had four steps up to the top and then four steps down inside the circle, where there was a large pit. Beside the pit were tall perches for staff to record trades among the dozens of brokers simultaneously yelling and confirming deals. With buyers and sellers moving around, raising and lowering hands, and shouting bids loudly, it was chaos.

In the early years, each of the eight sides represented commodities, and the steps were supposed to indicate which month the broker was trading in. Moving from section to section in the pit became difficult, so traders would include the commodity they were buying or selling in a verbal outcry. A hand motion towards oneself indicated an interest in buying and a hand motion away meant the broker was selling. A broker might be yelling, "1/4 for 1,000 May . . . 1/2 for 1,000 May," indicating 1/4- or 1/2-cent increments of buying interest for 1,000 bushels of canola (assuming they were in the canola pit). The seller would say "May at 3/4s . . . May at 1/2." Anyone trading would have to be in the pit or on the top platform.[40]

The market was only open for a specified time (9:00 a.m. to 1:00 p.m.) each business day with all buyers and sellers present. Before electronic boards there would be chalkboards all around the walls of the trading room, with notations continually being manually changed by personnel, recording the last trading price of each commodity according to bids and offers in the pit.

---

40   Explanation of pit trading courtesy of Russ Crawford.

Traders would come to work early in the day to review news and events that had occurred overnight that could affect trade that day. Then after four hours of fast-moving, high-adrenaline, and stress-filled trading, the traders were ready for a release. The Lock, Stock and Barrel was a pub next to the old Grain Exchange Building where traders would gather for beer and lunch, in that order. The beer was cheap and there was a popcorn machine that dispensed fresh, salty popcorn for free.

After noisily reviewing the day's events on the trading floor, satisfying themselves with lunch and no small amount of beer, many traders went straight home. I have witnessed this daily event. Unfortunately, the pub burned down in the 1990s.

The Exchange is now all electronic and is owned by an American company, International Commodity Exchange, known as ICE Futures. Physically trading grain in a pit is now a lost art and a colourful memory of the past.

One of the issues that stands out in my time on the WCE board was when XCan Grain attempted to profit by cornering the canola market. XCan was the international trading arm of the three Prairie Pools and traded all non-board crops for the Pools. Xcan was one of the largest traders on the WCE. When the situation of a potential market squeeze occurred, members of the WCE asked the board of governors to step in and have events leading up to the situation reviewed. The investigation determined that Xcan had acted in a manner that resulted in them profiting from cornering the market. Xcan protested loudly, and the investigation became a long-drawn-out affair, with Xcan eventually paying $250,000, ostensibly to cover the investigative costs of the Exchange, with no admission of guilt. [41]

Ironically, when the Pools were established in the 1920s, they would not use the Winnipeg Grain Exchange (as it was then called) to hedge their grain trading since they philosophically opposed it. This lack of risk management in 1929–1930 contributed to the bankruptcy of the three Prairie Pools in 1930. Now, sixty-five years

---

41 Explanation of the event courtesy of John De Pape.

later, they did the same thing that they accused the grain trade of doing back then -- manipulating the market!

THE RACE IS ON

Rationalization and investment in the country elevator system occurred rapidly after the WGTA was repealed and the system started to consolidate. All the line elevator companies began to compete to find sites for new construction while abandoning the old wooden elevators. New entrants such as French grain merchant Louis Dreyfus began investing in country facilities. In many locations groups of farmers joined together, raising capital and building large, farmer-owned, modern terminal collection facilities with large railcar spots of one hundred cars or more. These included the Prairie West Grain Terminal at Dodsland and North East Terminal at Wadena, both in Saskatchewan.

In 1985 there were 1,967 operating units (locations) purchasing grain on the Prairies. By 1996, ten years later, the number was down to 1,340, and by 2003 there were only 425. Farmers would have to truck their grain further, but elevator companies were able to negotiate preferential rail rates for the large car spots and pass these savings on to farmers. The nature of the business had companies competing for farmer's grain. The newer and larger grain elevators could accommodate the large semi-trailers that farmers were now using to haul grain from their farms. Farmers were consolidating their smaller acre farms into larger, commercial businesses.

In 1996 SWP followed UGG's lead and sold shares to the public to deal with their issues of paying out their patronage dividends and raising additional capital for new elevator construction and business expansion. Unlike UGG, SWP did not change their governance structure to seek experienced non-farmer board members to enhance the governance of the company.

With the changing market conditions, other grain companies were also looking around at potential mergers or new operating arrangements that would assist them to better position themselves in the new commercial environment. UGG reached out to each of the Pools to discuss possible business arrangements, but the Pools were not interested. Later, it was learned that the three Pools were in merger talks, but those discussions did not result in any new alignments, either.

The industry always wondered why the three Pools didn't merge. That event would have created a massive organization controlling the flow of grain on the prairies with 65% of the business. Apparently, Saskatchewan Wheat Pool, being the largest of the three, was always more interested in taking over rather than merging and that was not acceptable to either Manitoba Pool Elevators or the Alberta Wheat Pool. There likely would have been issues raised by the Competition Bureau as well.

This situation was even more pronounced in more recent merger talks. Don Loewen, the aggressive new CEO of SWP, was in a strong financial position, having just sold shares to the public. AWP and MPE did not have access to additional cash and the unpaid patronage dividends were hanging over their balance sheets like dark clouds. In this environment, the only acceptable deal to SWP was a complete takeover of MPE and AWP. Once again, this was not acceptable to either MPE or AWP.

Loewen at one time had approached Allen from UGG to meet and discuss some sort of arrangement. It was a clandestine meeting, with just the two of them present. However, there never was any real type of discussion. Loewen aggressively attempted to verbally bully Allen into a deal, which was essentially an SWP takeover. The meeting did not last long, and Allen flew back to Winnipeg.

In 1997, the Alberta Wheat Pool (AWP) and Manitoba Pool Elevators (MPE) jointly attempted a rather sloppy and ill-conceived hostile takeover of UGG. Earlier they had individually refused to enter any kind of preliminary discussion with UGG about a new business arrangement. Instead, they stealthily plotted together in the background to carve up and share the UGG assets. A hostile takeover event was the plan. According to Russ Crawford, who worked at AWP at the time, "The plan blew up in the group's face when AWP's brokers inadvertently acquired more than 10% of UGG's shares and triggered a public declaration of intent." Once UGG knew of this, they were able to mount a defence against the hostile action.

At this point, the balance sheets of AWP and MPE were very weak. Unlike UGG and SWP, they had not dealt with their issue of unpaid patronage dividends.

Once a takeover action is initiated, you cannot take defensive action just because you don't like the suitor. As directors, it was our duty to see if we could find something better for our shareholders. Going back to the old arrangement was not an option. Board members must always act "in the best interest of the corporation," meaning shareholders. The UGG board did not feel the Pools' offer of $14.00/share was sufficient and believed the two Pools were too weak financially to benefit UGG shareholders.

UGG had a good relationship with Archer Daniels Midland (ADM), an American agri-food company. The two companies shared an investment in a canola-crushing plant in Lloydminster, Alberta, and at times had explored the idea of ADM investing some money in UGG.

A deal was struck between ADM and UGG whereby ADM purchased 45% of UGG shares at $16.00/share. ADM received two seats on the UGG board. This allotment of directors had to come from the three non-farmer directors, so Art Mauro and Jon Grant stepped down. Two senior executives from ADM, Charlie Bayless and Bernie Kraft, joined the UGG board.

AWP and MPE went on to merge and rebrand as Agricore in 1998. They tried to keep up with investments in new facilities like other companies were making but their weak balance sheet and lack of access to new capital hampered them. By 2001 they were insolvent.

≈

Meanwhile at Saskatchewan Wheat Pool, after going public, CEO Loewen went on a spending spree. SWP invested in high-risk ventures such as grain terminals in Poland and Mexico, and non-core businesses such as a fish farms and pork production facilities. SWP launched the ambitious Project Horizon, which included building twenty-two large new terminal facilities in all three provinces under the name of AgPro Grain, a fully owned subsidiary of SWP.

Traditionally SWP farmer members were ideologically opposed to the construction of large inland terminals such as those that AgPro was building. Because they built the new facilities in the three Prairie provinces under the name of AgPro, many farmers, including SWP members, did not know this was a subsidiary of SWP.

In 2000, there were fourteen new facilities in three provinces licensed under AgPro Grain with an average size of 39,000 tonnes. This violated a long-time agreement between the Prairie Pools, whereby each would only construct facilities in their respective province and not invade the other Pool territories. SWP debt ballooned from $97 million in 1996 to $540 million in 1999. Things were unfolding in a manner that made it apparent the board of directors of SWP had lost control of their CEO, Don Loewen.

With these aggressive actions, SWP began to lose the confidence of its historically loyal base of Saskatchewan farmers. SWP's handling share of grain in Saskatchewan fell from 61% in 1993 to 33% in 2003. Reckless management of SWP was the main factor in its financial downfall. SWP went public in 1996 with a share value of $14.00. In 1998, the share value was $25.00. By 2003 the share value was 18 cents. Saskatchewan farmers who spent all their lives doing business with SWP to earn patronage dividends for their retirement essentially lost it all.

The SWP board finally realized their situation in 1999 and dismissed their CEO, Don Loewen. He was replaced by Mayo Schmidt, formerly with Conagra Grain in the US. Over the next few years, Schmidt would prove to be a brilliant tactician and formidable competitor who was underestimated by his rivals in the industry. He crafted and oversaw the execution of a totally unexpected and unpredictable turn of events in 2006.

So, at the beginning of the new millennium the two merged Prairie Pools (Agricore) were facing insolvency and the third (SWP) was severely impaired financially. UGG's early action to go public, take on outside directors, and be somewhat more disciplined in new investments paid off. UGG had a strong owner-partner in ADM. It looked like UGG was doing well in the Prairie farmer–owned grain-handling business. UGG had some preliminary talks with Richardson Grain about looking at a combined arrangement, but this did not go anywhere. Richardson was only interested in a deal whereby they would take control over the company and the UGG board was not agreeable to this arrangement. But things were not over yet.

> Overly optimistic business projections, under capitalization and undisciplined investments drained the finances of the Prairie Pool grain companies.

During this period in the mid- to late-1990s, all grain companies were rapidly building new, large terminal elevators in the country. Every time a new facility was proposed, management would bring to the board a business plan reflecting the potential profitability of the new facility. Each company would assume that when a new terminal was built their market share would rise in that area and it would be a good investment. However, all companies were doing the same thing. The amount of grain to handle was finite. This was a zero-sum game. Competing companies were each claiming the same potential new business at new construction points. Essentially, the grain available for delivery was being double counted. Each company wanted to develop a business plan that forecast profitability of each new build. The farmer-dominated boards of the farmer-owned companies were disregarding this reality of double-counting available grain because each wanted to see more new facilities built under their banner.

IN A LAND OF THE BLIND THE ONE-EYED MAN IS KING

Early in 2001, UGG received a surprising and unexpected phone call from Agricore, the new company that had been formed with the merger of the Alberta Wheat Pool and Manitoba Pool Elevators. Agricore was in dire financial distress, and their indebtedness would be called if they did not come up with some new arrangement or cash injection. They were interested in merging. By now, UGG, though the smaller company, was much further along than Agricore in consolidation of its country system and construction of large, modern facilities. UGG had a strong partner and investor in Archer Daniels Midland (ADM)0. and was financially stable. Unlike what UGG had accomplished in 1992, Agricore had not dealt with the issue of patronage dividend payouts to its members, and this left them with a weak balance sheet.

Last UGG Board prior to merger with Agricore

United Grain Growers
Board of Directors 2000-2001

Seated left to right: Craig Hamlin (ADM), Terry Youzwa, Ken Motiuk, Ted Allen,
Wayne Drul, Ernie Sirski, Brett Halstead
Standing left to right: Maurice Lemay, Spence Sutter, Henry Penner, Bernie Mackay, Jeff
Neilson, Gordon Cresswell, Bill Woodward (non-farmer), missing Bernie Kraft (ADM)

Members of the UGG board and management team were almost
giddy with delight. For years SWP had dominated the industry. All
other companies lived in the shadow of SWP, and the CWB/Pool
alignment had dominated the industry. SWP was also in financial
trouble, and the new Agricore was asking UGG to take over/merge
and bail them out. UGG had made it to the apex of the group of
farmer-owned grain companies. A merger of UGG and Agricore
would make the new combined entity (essentially UGG) the largest

grain handler on the Prairies. It had triumphed over the Pool grain co-ops . . . so far.

At the turn of the millennium, Ted Allen, chairman and president; Bryan Perkins, Alberta VP; and Roy Piper, Saskatchewan VP were formally and informally the most senior on the board. These three made a good leadership triumvirate. They were close to the same age, had been on the board together for a long time, and shared a similar business philosophy. Together they had shepherded the board through the last few years when we were positioning UGG for a new competitive environment. Perkins and Piper were well liked and respected by other board members and either would have been capable of taking over the chair. However, in 2000, Perkins retired from the board, soon to be followed by Piper in 2001. I took over the position of Alberta VP and Terry Youzwa took over as Saskatchewan VP.

The karma around the board table shifted when directors Perkins and Piper left. Chairman Allen was longer tenured than all the other farmer directors on the board, having been first elected in 1973. The next longest tenured were Youzwa and me, having both been elected in 1990. Allen was older than all the other farmer board members. Rather than working with the board, Allen began to consult more with CEO Brian Hayward. It appeared that more and more decisions that came to the board table had already been discussed and decided upon by Allen and Hayward.

It was after Perkins and Piper left that the board began to function less like a board should. Allen did not like to be questioned or challenged, and some board members began to agree with whatever decisions he and Hayward made without meaningful debate around the board table. These actions weakened the governance of the board. The chairman and CEO were making decisions and board members were having difficulty questioning and having full discussions. Those with total loyalty to the chairman would not challenge him and blindly supported him. One member of this group later would have the dubious distinction of being board chair when SWP took over Agricore United in 2007. At least one other ex-UGG board member was harbouring aspirations to be chairman as well, but he did not make it to the chair in the 2002–2007 period.

This evolution of board functionality was the exact opposite of the enlightening trends in corporate governance that were emerging and strengthening at this time. No efforts were being made to groom a successor. Perkins or Piper would have been clear choices to step into the chair position at any time, but they were now gone. An atmosphere resembling that of sibling rivalry was developing around the board table.

This was how the UGG board was functioning when Agricore came to UGG with a merger proposal in mind. The general feeling among senior management and the board was that UGG was now poised and positioned to totally lead the industry. UGG just had to go through the motions of the merger. Adrenaline on the UGG side was flowing freely so the subtle changes in board dynamics in the background faded in the overriding excitement of the merger process.

The name Agricore United (AU) was chosen for the newly merged company. The merger had to move quickly because Agricore had tripped financial caveats that had to be dealt with as soon as possible. There was little serious pushback from the Agricore board. The transaction was a takeover in every aspect but name. UGG moved carefully so as not to unsettle Agricore and its board and personnel in the process. UGG wanted the customers of Agricore to deal with the newly merged company. It was agreed that UGG's Allen would remain chairman and Hayward would remain CEO of the new company. Most senior executives; the corporate secretary; the information technology and human resources staffs; and the chief financial officer would all be from UGG. However, the boards of UGG and Agricore had to merge in some manner.

The new company was to operate under the same charter as the old UGG. The governance structure would be the same as that of UGG. If we wanted something different, we would have had to go to Parliament to change the UGG Act and that was a long slow process.

The new board was to be made up of twelve farmers and three non-farmers. It was decided that six farmer-directors from each of the parent companies would form the new board. Both UGG and Agricore boards went through an internal selection process as to who would go forth and who would retire. Retiring directors from UGG were paid $40,000 as severance. The six directors from UGG that went on to the newly merged company were Ted Allen, Wayne Drul, Maurice Lemay, Ernie Sirski, Terry Youzwa, and me. Joining the new board from Agricore were Mel McNaughton, Hugh Drake, Don Lunty, Rob Pettinger, Neil Silver, and Jim Wilson. The three non-farmer directors were Allan Andreas, CEO of ADM; Paul Mulhollem from ADM; and Bill Woodward, an experienced corporate director.

Of the five executive members of the board, four were ex-UGG. The executive included the chairman, vice-chairman, and three provincial vice-presidents.

An interesting sidebar in hindsight that became pivotal for me occurred during UGG's selection process for the six directors. The UGG charter called for three-year terms for elected directors. A problem arose between Ernie Sirski and me. Both of us had two years left in our terms, but this needed to change to keep future elections in sync. One of us would have to modify our remaining term so there would be a balance going forward with an equal number of directors to be elected each year in the future. One of us would have to stand for re-election in the upcoming annual meeting of 2002 rather than completing the original full term to 2003. It was felt that it would be too divisive to put this to a vote in such a small group, so we decided to simply flip a coin. Ernie won. This meant I would have to stand for re-election at the upcoming annual meeting in the fall of 2002 rather than complete my original term that was to expire in 2003.

Since the grain-handling industry was so concentrated and the new company would have such a large share of the Prairie grain handle, the merger had to be approved by the Competition Bureau. The Bureau did not wish to see concentration of company ownership in localized areas, which would essentially result in regional monopolies. So, upon review, the Bureau ordered divestiture of some facilities. This exposed the new company financially because the industry knew which facilities had to be disposed of before the merger would be allowed to proceed. This gave the proposed new company little time to negotiate good divestitures. In these deals UGG/Agricore was essentially negotiating with parties who knew which facilities had to be sold. And the clock was ticking towards the deadline of the bank covenants of Agricore. This entire exercise was financially damaging to the pro forma opening balance sheet of the new company, Agricore United.

In particular, the Bureau's ruling on facilities in the Port of Vancouver created a difficult situation. The Port of Vancouver is very crowded. Agricore had a newer terminal on the east end of the south shore, and then as you moved westward there was a UGG facility and then an old Pacific Grain facility owned by Agricore. The UGG and Pacific Grain facilities were older and smaller and in a crowded and awkward location to spot loaded railcars of grain. The UGG facility had low draft at its docking facility as, apparently, there was rock that would not allow for further dredging to allow larger vessels to complete loading.

The Competition Bureau would not allow the merger to proceed unless one of these facilities was divested within an allotted time. The gem of the terminals was the newer and larger Agricore facility. The Pacific facility had lots of storage, so the UGG terminal was chosen to be sold. The problem was the entire industry knew about this situation and there were not many buyers in the first place. This was like playing poker with all your cards laid out for your opponents to see.

The UGG facility was eventually sold to Alliance Grain, a consortium of smaller, privately owned companies which over time included Patterson Grain, Providence Grain, P&H and North West Terminal. The facility was first licensed in this name in 2007, five years after the Competition Bureau ruled it must be sold.

As the summer of 2002 unfolded, the Prairies were having a dry summer, and a short crop was widely forecast. To make the numbers work, the pro forma budgets for the newly merged company were aggressive and a good crop volume was needed to meet the handling projections. A short crop would be disastrous. Other events were also occurring in the summer of 2002 that would later shape the future. One of these had its roots at the UGG board table before the merger.

Company management had been pushing the board for some time to address the cost of farmer involvement in the company's democratic process. There were the costs of the local members advisory groups from the country. There was a large cost of hosting and bringing all the farmer delegates to the annual meeting. There was the cost of the company staff required to administer this. Management pointed out that all these were costs that UGG faced while its competitors did not.

Also, in the days when the Pools and UGG operated as co-ops and quasi-farm organizations, the position of president and chairman of the board was considered a full-time job with a handsome salary. As UGG became a public company and its role in farm policy greatly diminished, it was felt that the chairman position should become a part-time position with a lesser salary as in other public companies. Diminishing this role also fit within the new governance standards other companies were adopting—that of the separation of board and management. It is always difficult to cut a position back, so we had to tread lightly.

An opening arose when Allen asked if he could move to Alberta. The board agreed, provided he would scale back his duties to part-time, with a corresponding salary. Parties agreed in principle but when Allen moved to Alberta he would not relinquish his full-time salary. I was chair of the Compensation Committee at the time, so we put this on our agenda. As well, human resources presented the committee with a billing Allen had submitted for his moving expenses, which should have been Allen's personal expense.

This became problematic since the cost of the move was never discussed between Allen and the board. According to company policy regarding employee moves, reimbursement depended on which party initiated the move. If the company initiated the move, the company would pay the costs. If the employee initiated the move, the employee would pay the costs. Provision was made for exemptions to the policy if the employee requested expenses paid under special circumstances. Regarding Allen's move, he asked to relocate and did not request or obtain prior approval to have his expenses paid.

The Compensation Committee agreed that according to policy Allen did not qualify to be reimbursed for moving expenses. The committee also decided it was time to clarify the issue surrounding the reduction in the salary of the chairman as per the original verbal agreement on his relocation to Alberta. In a highly unusual turn of events, the Compensation Committee recommended both to the board, and the board then overturned the recommendations.

Several board members felt this was inappropriate conduct for the chairman. This reflected the dynamics of the board at this time, and the way Allen worked with the board. It also exposed the number of board members who sought his approval and would not take any stance to challenge him. The role of board members is to act in the best interests of the company, not blindly support the chairman. It is doubtful Allen would have tried something like this when Perkins and Piper were around. Allen's moving expenses were eventually paid by the company, but the issue of his salary became caught up in the hubris surrounding the merger and the issue later resolved itself when he left the chair.

In the early summer of 2002, Allen requested a leave of absence for a medical procedure. No plans were put in place to discharge his duties and commitments during this time. Company VPs were not aware of what was required of them in his absence and kept getting last-minute phone calls from Allen's office to fill in for his prior obligations.

≈

As the summer of 2002 turned into fall and we approached the first annual meeting of the merger, it was known the Prairies had a short crop in 2002. The newly merged company would not be able to reach its handling targets in year one of the merger. Some directors, including myself, were concerned about this and requested revised pro forma budgets based on the lower crop supply from the drought-stricken 2002 crop. Management did not provide these revised budgets. Another faction of the board, led by Allen, wanted the transaction to proceed without calculating the new numbers. Revised pro forma budgets based on the short 2002 crop were not prepared before we went into the annual meeting in February of 2003, with the board recommending the merger to the combined delegate body without providing them with the updated financial information.

The chair, CEO, and some directors were so caught up in the exuberance of the takeover and subsequent dominance of Agricore United in the industry that they did not want to see any budgets or projections that would rain on their parade. The merger went ahead, but the board and management did not acknowledge the magnitude of the evolving financial challenge the new company was about to face. The three non-farmer directors were quite concerned about the upcoming year. I contacted Allen several times that summer to express my concerns about the potential short crop and its effect on the pro forma budgets of the newly merged company. His acerbic reply was, "Quit looking outside your window."

As it turned out, due to the drought, the 2002 crop on the Prairies was down 38% from the five-year average. Heading into the merger meeting in February of 2003, management and many members of the board were not preparing for this reality. Repeatedly we heard statements from senior management to the effect that Agricore United (AU) is now the largest grain handler on the Prairies and the new AU will be just fine. Some board members shared the same belief. It was to be a tough start for a company that was volume-driven, highly leveraged, and only marginally profitable at the best of times, not to mention it was still in the process of merging two cultures.

First combined UGG/Agricore Board – 2002

Left to right standing: Maurice Lemay, Hugh Drake, Ernie Sirski, Don Lunty, Paul Mulhollem (ADM), Jim Wilson, Rob Pettinger, Neil Silver.
Left to right sitting: Bill Woodward (non-farmer), Terry Youzwa (Saskatchewan VP), Wayne Druhl (Manitoba VP), Ted Allen (chairman and president), Mel McNaughton (first VP), Ken Motiuk (Alberta VP), Allen Andreas (ADM).

The ninety-fifth annual meeting of UGG was held at the Fairmont Hotel in Winnipeg from February 5 to 7, 2003. This was two months later than the usual November date for an annual meeting due to the amount of preparation required to complete the merger. Because of the merger, this would be one of the most historic annual meetings in the history of UGG. Not since the Grain Growers' Grain Company amalgamated with the Alberta Farmers' Co-operative Elevator Company in 1917 to form UGG was there a meeting of such importance to the company.

As we were going into the annual meeting in November 2002, I was receiving support and encouragement from several sitting board members to let my name stand for chairman at the board reorganization meeting following the annual meeting. This meeting was where the board members voted for the chairman and VPs for the upcoming year. I had the support of the non-farmer directors (including the two ADM representatives), and somewhat ironically, the support of ex-Agricore directors. In January of 2003, I had met with Bill Woodward, the non-farmer director, and Neil Silver, previous chairman of the board of Agricore. They both strongly encouraged me to stand for chairman. Ex-Agricore directors felt they were being treated condescendingly and without respect by the new board chaired by Allen.

I did not know the ambitions of the other ex-UGG directors. Even though we were congenial to each other, it was likely others harboured aspirations to become chairman when Allen eventually left. Even without their support, I felt I had enough support to become chairman at the board meeting after the annual meeting. But first I had to get re-elected by the delegate body.

That year's delegate body was the first with a combined group of delegates from the former UGG and the former Agricore. Previously I had strong support from UGG delegates, having gone unchallenged in my last two re-elections. The Agricore delegates did not know me as well. The events prior to the election for director at that annual meeting took an unusual turn and information was conveyed to me later by delegates who supported me and were surprised at what was happening in the back rooms. It appeared Chairman Allen was fearful that I may challenge the chairmanship. The only way to stop me from challenging his chairmanship was to ensure I did not get re-elected by the delegate body and get back to the board table.

Paul Earl writes: "One of the prime candidates to succeed Allen was Ken Motiuk, and he was one of those who believed that change was needed. He farmed in Mundare, Alberta and had been on the board for over a decade. He began to challenge Allen and made it known that he would seek the presidency after the 2002 annual meeting. However, his term as director was expiring, and so he first had to be re-elected to the board by the delegates. The dispute became public at the annual meeting. Allen was not anxious for Motiuk to rejoin the board and sought to influence delegates to vote against him."[42]

With the assistance of UGG Member Services Representative Glynnis Perkins, a smear campaign was conducted against me on the evening before the directors' election. Derogatory and disparaging falsehoods were spread about my character and suitability for re-election as director. I was informed of this by delegates who attended the small meetings that occurred in hotel rooms.

This was ironic because previously at the board table Allen would remind sitting directors not to get involved in the election of other directors, and certainly not to include UGG staff in the election process. He would go on to say that if any directors became involved in another director's election they would be asked to resign from the board. Employees who participated in a director's election would be let go from the company. And then he went on to do exactly what he told other directors never to do, while engaging company personnel in the process. Glynnis Perkins stayed on with the company after the 2002 annual meeting and Allen stayed on the board as a director.

I cannot be sure if other ex-UGG directors were aware of what was going on that evening, but I do know some of them were reluctant to challenge Allen's actions since my removal would open the way for one of them to take over the chair after Allen retired.

During the election the next day, I sensed something was going wrong because my re-election was going on for more than one ballot.[43] In the past three elections I was either acclaimed or I won on the first ballot. This election went to five ballots. The result was that I was not re-elected to the board of directors by the delegate body.

I immediately went to one of the non-farmer directors whom I trusted and told him what had just occurred.

---

42 Earl, *The Rise and Fall of United Grain Growers*, page 171.

43 The UGG director election required a 50%-plus-one vote from the delegate body; simply receiving the most votes did not result in a victory. When there were multiple candidates, should no candidate receive 50% plus one of the votes cast, the candidate with the fewest votes would drop off and the delegate body would go to another ballot.

It was customary that the chairman would briefly address the delegates at the conclusion of the election and call for a motion to adjourn the annual meeting. That year there was a bit of a delay before Allen came to the podium for his closing address to the delegate body. In a totally unexpected announcement, Allen announced he would not seek the chairmanship for the upcoming year.

Prior to this, Allen had given no indication that he was intending to vacate the chair. To the contrary, Allen had previously let the board know he wanted to stay on as chairman through the company's centennial in 2006. The board did not have a succession plan for the chairman. Now, after his shenanigans during the election, I believe some directors went to Allen prior to his closing remarks and told him in no uncertain terms that he was to immediately resign as chairman.

The usual procedure at UGG was that immediately after the conclusion of the annual meeting, the board would meet and the president and chairman of the board for the upcoming year would be elected by the new board members. The company vice-president and provincial vice presidents would be elected at this time as well. This board reorganization meeting was usually short and ended with a group dinner. Apparently, following this fiasco, the reorganization meeting lasted several hours. I do not know what was discussed. I was not there. My understanding is they did not go to dinner together as a group that night. It appeared the board was unsettled and divided.

John Jacob Aster III once said, "Always take the trick. When the opportunity you seek is before you, seize it. Do not wait until tomorrow on the supposition that your chance will become better, for you may never see it again." This time it didn't work!

We spent an unsettling evening in our room at the Fairmont that night and made many phone calls to friends and colleagues to share with them what had just happened. Only one director, Ernie Sirksi, called to see how I was doing.

Two tigers cannot share the same mountain.

Chinese proverb

THE AFTERMATH

And just like that it was all over. One day you are thinking about the company and planning for the future, prioritizing tasks and assessing personnel. And the next day you are at home on the farm looking at an empty calendar. The former chairman's actions were unprecedented and unprofessional, and it was clear the board was now in disarray. I considered legal action against Allen and the company for what had just occurred, although it was a long shot. I decided it just wasn't worth it.

I was hurt and disappointed by the actions of some of the UGG directors whom I thought were above condoning these actions by Allen. It was clear they were afraid to challenge him even when he went far beyond the bounds of acceptable protocol for the chairman, as established by himself.

In the summer of 2003, along with legal counsel, I met with Agricore United (AU) Chairman Jim Wilson and AU in-house legal counsel Chris Martin in Edmonton. After some negotiation, I agreed to a $40,000 settlement. This was the amount other retiring UGG directors received when they resigned to make room for Agricore directors on the newly merged board.

At this point, the legacy UGG looked like the rising star with AWP and MPE now wrapped into one company, Agricore United, led mainly by ex-UGG personnel. Saskatchewan Wheat Pool was in a dire financial situation and its survival was in question. But the first years were difficult for the fledgling company. The short crop of 2002 hurt Agricore United in its first year of operation. Prairie grain production in 2002 was only 62% of the previous five-year average. In its first three years of existence, AU, even though it was the largest grain-handling company on the Prairies, lost money every year—to a cumulative total of $33 million. Clearly the new company was struggling financially.

In 2004 I wrote an article, "The Decline of the Prairie Grain Co-ops,"[44] which was published in the January 17, 2005, issue of *Grainews*. In it I outlined the difficult financial position both AU and SWP were in. I didn't think AU would survive, and I thought perhaps it would be taken over by ADM. I accurately predicted the takeover, just not the company that would accomplish it.

A description of what happened to the new AU over the next four years is well documented in *The Rise and Fall of United Grain Growers* by Paul Earl. By 2007, 101 years after its inception and 5 years after the UGG/Agricore merger, the roots of the old UGG would no longer exist. It would be the last of the once powerful farmer-owned grain co-ops, which had included UGG and the three provincial Wheat Pools. At one time these four entities handled 75% of the grain on the Prairies. UGG was the first farmer-owned company to be established with roots going back to 1906, and the last to be taken over by the newly reconfigured public company, Saskatchewan Wheat Pool, in 2007.

The Prairie farmer–owned grain companies/co-ops collapsed due to weak corporate-governance practices and a lack of visionary leadership and business skills on their boards of directors. Boards of large companies everywhere look for people with specific skill sets to oversee their businesses. AU and SWP continued to elect farmer-directors with limited corporate-governance and financial skills. The SWP board in the late '90s did not have the business confidence to question and curtail the high-risk and unsuitable investment decisions of CEO Don Loewen; that nearly bankrupted them. The farmer-directors of Alberta Wheat Pool and Manitoba

---

44   Appendix #7

Pool Elevators failed to address their weak long-term capital structure and were now under a UGG-dominated governance-and-management group.

Unfortunately, after the merger, the Agricore United board did not amend their corporate-governance structure to provide for the appointment of additional skilled non-farmer–directors. This would have required an amendment to the UGG Act, but it might have been worth it. The ineffective leadership of the AU board during its brief tenure later resulted in a takeover by a newly invigorated Saskatchewan Wheat Pool in 2007. The weakness of the AU's farmer-dominated board became clear in the takeover by SWP.[45]

AU director Paul Orsak summed it up the best. "I think we could have positioned ourselves to become the aggressor rather than the acquired."[46]

The issue came down to the simple fact that farmer-directors were elected by farmer-delegates. Each election became a popularity contest rather than a search for skilled businesspeople for the job. In their quest to have delegates vote for them, candidates for director often focused their election platforms on very local issues rather than adopting a global view of what was in the best interests of the company long-term. One simply must witness the calibre of politicians that currently lead this country (in 2024) to see how well these types of elections work when it comes to managing financial issues. Adhering to this process was fatal for the Prairie grain companies/co-ops that were owned and controlled by farmer-customers.

---

45 A well-documented account of the last years of UGG is published in: Earl, *The Rise and Fall of United Grain Growers*, Chapter 8.

46 Earl, *The Rise and Fall of United Grain Growers*, page 177.

A PHOENIX RISES

At the time of the UGG/Agricore merger in 2002, SWP was totally mired in the unmanageable debt and hubris left behind by former CEO Don Loewen. Newly appointed CEO Mayo Schmidt started to aggressively cut costs and divest from non-core assets. He then utilized a plan he called fresh-start accounting and revalued all assets at 2003 levels. Schmidt engineered a $405-million debt-restructuring plan and creditors, bondholders, and shareholders all took a big write-down as hundreds of millions of dollars of unsustainable debt was written off or lost by shareholders. Assets were written down to pennies on the dollar. The creditors were in control and went with Schmidt's plan as they felt this reorganization was the best way to clean up the unmanageable debt hanging over SWP. They felt it was better for them to try to forge ahead and get pennies on a dollar rather than lose everything. Shareholders, including many farmers in Saskatchewan, witnessed their SWP shares become essentially worthless, but SWP was back in business.

In 2005, SWP became a full federal corporation under the Canada Business Corporations Act. They would no longer be governed by a farmer-elected board. They now had board members from the business community who were sought out for their corporate business skills. The once powerful and proud SWP that had dominated the Western grain-handling industry for many decades was now a fully public company and no longer a co-op.

Schmidt then positioned SWP to challenge the disarray at the new Agricore United. What was to come next was a bold and masterful play that took the industry by surprise. The wily Schmidt played the structure of the country grain–collection system like a chessboard. In a large transaction such as this, the Competition Bureau is concerned that competition would become limited in some localized areas. Knowing he would need Competition Bureau approval for his plans, the outcome made it clear he was proactively working quietly with them in advance to ensure they would approve his plan. He didn't want to face the same situation UGG had found itself in during the 2002 deal with Agricore whereby everyone knew what facilities Agricore United would have to divest themselves of, leaving AU in a weak bargaining position for the facilities they had to sell.

By working behind the scenes with other grain companies, Schmidt negotiated facility swaps and/or purchases and divestitures to appease Bureau concerns. This reshuffling of assets included line elevators in the country, farm-supply outlets, and terminal facilities in the Port of Vancouver.

In October 2006, SWP, under the leadership of Schmidt, having barely reached some semblance of financial stability, made an unsolicited public offer for all the shares of AU. At this time, AU was a much larger company with a stronger balance sheet. The bold brashness of this plan shocked the industry.

The idea that SWP would take over AU was not at all appealing to the AU board. However, once an action like this is launched, the world never goes back to what it was. AU could not reject SWP bids just because they opposed the action. Their fiduciary duty as directors compelled them to act in the best interests of the company with the interests of shareholders being paramount. If they didn't like the SWP offer, they were compelled to come up with something better. AU approached James Richardson International (JRI) to devise a counter proposal that could be presented to shareholders as an alternative to the SWP offer. A bidding war between SWP and JRI ensued.

The coup de grâce came in May of 2007 when Schmidt went to JRI and offered to sell them several AU assets should SWP be successful in the takeover. JRI later purchased fifteen AU country terminal facilities and nine of AU's crop input outlets. JRI was now in a win-win situation and decided to accept SWP's offer. Schmidt took the extra cash and made a new, and what was to become a final, bid for AU. JRI, apparently agreeable to the new metrics of the deal, did not submit a higher bid. AU was compelled to sell to SWP. Schmidt had carved up AU like a Thanksgiving turkey and shared the spoils with JRI and others. Several swaps and sales/purchases then occurred between existing grain companies, allowing them each to strengthen their country networks. Terminals in Vancouver changed ownership as well.

On April 19 and 20, 2007, I was taking Module 1 of the Directors' College Charter Director Program offered by Dalhousie University. One of the guest speakers was Jon Grant, the Agricore United director chairing the board committee dealing with the SWP takeover. At dinner Jon came by and asked if he could join me. We had a very interesting discussion.

The once strong and proud farmer-owned Prairie grain companies, the three Wheat Pools, and UGG were now a part of history.[47] SWP, having essentially gone bankrupt eight years before, once again became the largest grain-handling company on the Prairies. It was no longer a farmer co-op—it was now a public company. Schmidt went on to rebrand SWP as Viterra, which was later purchased by global conglomerate Glencore.

In his management book *From Good to Great* Jim Collins states that "good is the enemy of great."[48] Most people are satisfied when they get good, so they stop trying to become great. Being great calls for working harder and not resting on your laurels. In the period of 2003 to 2006, the AU board and management were satisfied with being good and thought they had command of the industry

---

47 These events are all well documented in *The Rise and Fall of United Grain Growers* by Paul Earl, Chapters 9–13.

48 Collins, Jim. *From Good to Great*. HarperCollins Publishers. 2001. pg. 1.

because of their dominating size. But Mayo Schmidt wanted SWP to be great. He was relentless in his pursuit of taking over AU and he succeeded.

In the battle for control, AU did not show the energy, desire, zeal, or inventiveness that UGG had set in motion to fight off the Alberta Wheat Pool and Manitoba Pool Elevators' takeover attempt in 1997. There did not appear to be any aggressive leadership at the board or senior management level to fire up the troops as had occurred in 1997. The chaos and instability on the board precipitated by Allen's actions at the annual meeting in 2003, and the resulting lack of leadership and direction, was evident. Various factions had formed on the board, and unity of focus and action was not apparent. In four short years, the new Agricore United had squandered the opportunity to be not only the largest grain company on the Prairies but the only one controlled by farmers.

If one is to accept the premise that management's defense of AU against a SWP takeover was not sufficiently vigorous, perhaps one should investigate the size of the golden parachutes that members of AU's senior management were entitled to should the takeover proceed. There were stock options and severance packages at stake, and with a sale price of $20.25/share there was potentially a lot of money to be made personally by some members of the management team.

The 2004 AU Annual Report reports there were 732,045 stock options outstanding. The average exercise price was just over $10.00/share. At a sale price of over $20.00/share there was potentially over $7 million in profit on these options for the few members of the senior management team of AU that held them. Prior to the SWP's offer to purchase, AU shares had been trading just above $8.00/share. It was up to the board of directors to be mindful of these types of issues as they wove their way through the negotiations. This begs the following questions: Were some members of the senior management team more interested in cashing in on their stock options, which had just become quite valuable, rather than recommending what was really in the best interests of AU? Why was the AU board so passive as this was all evolving?

Number of mainstream grain-buying facilities on the Prairies

	1976/77	1990/91	2000/01	2004/05	2008/09	2023/24
Agricore	0	0	181	0	0	0
Agricore United	0	0	0	87	0	0
Alberta Pool	789	272	0	0	0	0
Cargill	253	99	45	33	40	25
G3	0	0	0	0	0	19
Grains Connect	0	0	0	0	0	4
Louis Dreyfus	0	0	10	10	10	0
Manitoba Pool	284	139	0	0	0	0
P & H	65	32	23	19	18	26
Patterson	89	52	48	46	35	21
Richardson Pioneer	443	201	75	62	63	55
Sask Pool	1317	476	133	48	0	0
UGG	701	285	79	0	0	0
Viterra	0	0	0	0	96	65
Total	3941	1556	594	305	262	215
Avg. size (tonnes)	2425	4365	9022	13,830	17,058	32,958

Source: Canadian Grain Commission, Grain Elevators in Canada.

The table above shows the mainstream grain-buying companies, the number of facilities each company had, and the average size of a country elevator. It reflects the massive consolidation and reconstruction that occurred over a forty-five-year period. From total domination in 1976, the farmer-owned grain companies and co-operatives no longer exist. Their remnants operate as Viterra, which is currently (2024) in the process of merging with Bunge Grain. The merger process is awaiting regulatory approval in the countries in which the new entity will operate.

Looking at the table we see the major shifts that occurred in the country elevator system from 1976 to 2020:

The farmer-owned companies, the three Pools, and UGG went from controlling 75% of the system facilities in 1976 to Viterra, a public company, controlling 30% of the system in 2023. The co-ops and farmer-owned companies are all gone. Richardson went from 11% of the system to 26% in the same period. Patterson and Parrish & Heimbecker combined went from 4% of the system in 1976 to 22% in 2023.

The companies that were the big winners in this consolidation and reconstruction period were the old, privately held, family-owned companies: P & H, Patterson, and Richardson. These companies were all in existence before the days of the CWB. During the period of the CWB and socialized wheat marketing, these companies all patiently laid low, as there were few favourable investment opportunities. Once the system deregulated, they methodically re-entered the system and are now large players and successful companies.

After deregulation, new entrants and foreign funds such as G3, (a partnership of Bunge and Saudi Agricultural and Livestock Investment Company, along with the Farmers Equity trust from the old CWB) along with Australian Grain Corp, began to invest in the grain-handling-and-marketing system. The first large, modern facility in decades was built in the Port of Vancouver by Bunge and began to operate in 2020 under the name of G3.

For a brief period, Agricore United had the opportunity to be the largest legacy company with farmer ownership. Now they no longer exist. All the work and dreams of having a successful farmer-owned grain company on the Prairies by the great leaders of UGG—Ed Partridge, Tom Crerar, John Brownlee, and Mac Runciman—were lost. The feckless and inept leadership of AU was outmaneuvered and outwitted by Mayo Schmidt and SWP in 2006.

The transportation system was now poised for the future. But we still had the CWB to deal with and that chapter begins in 2006 with the election of the Harper government.

EPITAPH OF A GRAIN CO-OP: SASKATCHEWAN WHEAT POOL

1923–1924—Two Saskatchewan farm groups, Saskatchewan Grain Growers Association and Saskatchewan Co-op Elevator Company, joined forces to form Saskatchewan Co-operative Wheat Producers (later Saskatchewan Wheat Pool). With a grand effort and in short order, they signed up over 50% of the wheat acres in Saskatchewan to be marketed in a pool. Together with the Pools in Alberta and Manitoba, the Central Selling Agency (CSA) was established to market all the wheat consigned to these pools from the three provinces. Saskatchewan, being the largest, led and heavily influenced operations of the CSA.

1924–1929—The world wheat market grew and world prices were strong. The Pools and the CSA were successful in marketing over 50% of Prairie wheat through a pool. They began to build elevators across the Prairies.

1929–1930—The wheat market crashed and the CSA found itself holding millions of bushels of wheat that had been purchased from farmers at a much higher price than the current. The CSA, in its ideological mistrust of the Grain Exchange, had not hedged these purchases. The CSA became insolvent, and the three provincial governments had to bail out their respective Wheat Pools.

1930–1943—The Wheat Pools operated as grain companies, handling farmers' grain much like private companies would and maintaining about 50% of the market share of grain handle on the Prairies.

1943–1960—During World War II, the CWB became mandatory, and increased control of system throughput was put into the hands of the CWB. This was exactly what the Pools had been advocating. During this period, the Pools, strong supporters of the CWB, saw their influence over system management grow as they worked closely with the CWB. Parallel to this was the Pools' growing influence politically over agricultural policy development in Western Canada.

1972—With the purchase of Federal Grain, the Pools had 65% of the grain handle on the Prairies. CWB railcar allocation policy served to protect the market share of the Pools by basing car allocation on past handling, thus creating a self-perpetuating cycle. Taxation policy for the Pools vs the private trade gave the Pools a financial advantage over other grain companies.

1970s and 1980s—The Pools, led by SWP, became powerful institutions in Prairie agribusiness and influential in policy development for Canadian agriculture. Financed by tariffs from grain handle, SWP became arguably the most influential farm lobby group in Canada. The SWP logo blanketed Saskatchewan with their multitude of grain elevators, livestock yards, and farm-supply warehouses. Almost every farmer at some point did some business with SWP. Upon making a small purchase such as a bundle of twine, a patron would be encouraged to purchase a $5.00 membership in SWP. Then, when speaking out on farm-policy issues, SWP would purport to have the support of all these members. The membership included many retired farmers or landlords on crop-share lease agreements. Membership lists were not always updated and purged, so they often included individuals who had passed away. This is why, at times, SWP membership could exceed the official number of farmers in the province.

1996—In order to pay out their members' equity as well as raise cash for growth of their business, SWP started to trade on the Toronto Stock Exchange to raise public funds.

1996–2000—Don Loewen was hired as CEO of SWP and went on a spending spree, taking on a financially unsustainable capital-expenditure program as well as spending money on non-core investments, some of them international. SWP took on more than it could handle and faced financial ruin. SWP share prices dropped from $24.00 in the late 1990s, to $0.30/share in 2003. Loewen was relieved of his duties as CEO in 1999 and Mayo Schmidt took over in 2000. SWP's share of grain handle in the province fell from 61% to 33%.

2005—SWP converted from a co-operative to a business corporation as defined by the Canada Business Corporation Act. After a stringent exercise of financial realignment, with both shareholders and bondholders taking a significant write-down and receiving pennies on the dollar, Schmidt kept SWP operating and launched a bold and unexpected hostile takeover of Agricore United. After successfully achieving the takeover, the roots of the four farmer-owned Prairie grain companies were united in one entity—albeit now a public company.

2008—SWP rebranded as Viterra, which was then taken over by international conglomerate Glencore.

2023—Bunge Grain proposes to merge with Viterra. The assets of the once powerful farmer-owned Prairie grain companies who once had 75% of market share would be buried in the vast capital structure of the international Bunge Grain that was one of the private industry villains in the earlier journey to build the Prairie-based, farmer-owned system over the past century.

The once unassailable holy grail of Prairie co-operatives, Saskatchewan Wheat Pool, would be no longer.

AS ONE DOOR CLOSES, ANOTHER OPENS

Country music artist Willie Nelson said, "Just when you're sure things are coming apart, they are actually just coming together."[49] Opportunities that were anticipated may disappear, but new and better ventures arise. In hindsight, I have to say this was certainly the case for me. Looking back at the chaos and stress that dominated the atmosphere on the board of Agricore United for its brief existence, along with the subsequent takeover of AU by SWP (Viterra), I no longer wish I had been there.

Meanwhile, all the while I was at UGG, we were growing our farm at Mundare, and we were seeding 4,400 acres by 2003. Our Motiuk farm received the Century Farm Award from the Province of Alberta. Homesteaded by my grandfather Ivan Motiuk in 1899, our farm has been farmed by our family for over one hundred years.

---

49 Nelson Willie. *It's a Long Story: My Life*. Back Bay Books. 2015. Page 94.

Congratulations & Best Wishes
To The Descendants Of The
IVAN MOTIUK & EVA(KULLY) MOTIUK FAMILY
On The Occasion Of The
-- 100th Anniversary --
NE 25-53-17-W4
ORIGINAL HOMESTEAD
1899 - 1999
Presented By
DEPUTY REEVE BILL J. ZELENY
COUNCILLOR, DIVISION 1
LAMONT COUNTY

The new initiatives we had introduced on the farm throughout the years were working well for us. One of the main hindrances was still the CWB with its quotas and price-pooling for wheat. This made it difficult to budget and forecast cash flow for wheat sales.

After the fiasco at Agricore United in February of 2003, I needed something more than the farm to keep myself occupied and upbeat. I found I had time on my hands, and I needed something to challenge my restlessness.

I initiated the Whitetail Crossing Golf Course and residential development at Mundare. We owned 240 acres of farmland on the southern boundary of the town, and its topography lent itself to the construction of a golf course. I took in two partners, one of whom owned an adjacent 360 acres of land, and another who had some experience in land development. Together we embarked on the Whitetail project, which included a golf course and housing development. We agreed on the name for the development, as well as the logo.

Golf architects Puddicombe and Associates were engaged to design a golf course. The town of Mundare annexed the property from Lamont County, and we were successful in having an area structure plan approved by the Town of Mundare. A golf course and development such as this is a risky venture and requires a great deal of capital.

In 2005 I sold my share to a Vegreville businessman, and the Whitetail project proceeded. Unfortunately, the financial crisis that hit in 2008 and the subsequent economic downturn stalled development. Currently, the real estate development component of Whitetail Crossing has been slow to progress. The golf course fared better and is now part of the Country Club Tour golf group based out of Edmonton. We now have a first-class golf course in Mundare.

At a display booth at the Edmonton Home and Garden show promoting Whitetail Crossing, circa 2005.

Throughout these years, I always stayed in touch with politics in the Alberta Conservative government. In 2002, I served on the Alberta Financial Management Commission. It was a small group of ten businesspeople from various sectors of the Alberta economy. Our assignment was to review Alberta's fiscal planning policies and strategies. Specifically, our task was to analyze, comment, and provide recommendations on:

- The Heritage Savings Trust Fund
- Infrastructure investment
- Debt management
- Alberta Investment Management Group

This experience provided me with a good insight into internal government workings and the entire fiscal situation of the Province of Alberta. I worked with some knowledgeable and influential people in the province and learned a great deal.

In 1999, I was appointed to the Alberta Economic Development Authority (AEDA), which was a small group of business leaders from the province who acted as an advisory body to Alberta's premier and finance minister. This was another opportunity to meet and exchange information with successful business leaders from across the province. My appointment stemmed from my working knowledge of the Prairie grain industry. Being a member of both AEDA and the Alberta Grain Commission provided me excellent access to the premier, cabinet members, and other industry leaders in Alberta. Through this access, I was able to brief many influential leaders on the detrimental effects of the CWB on the Alberta economy.

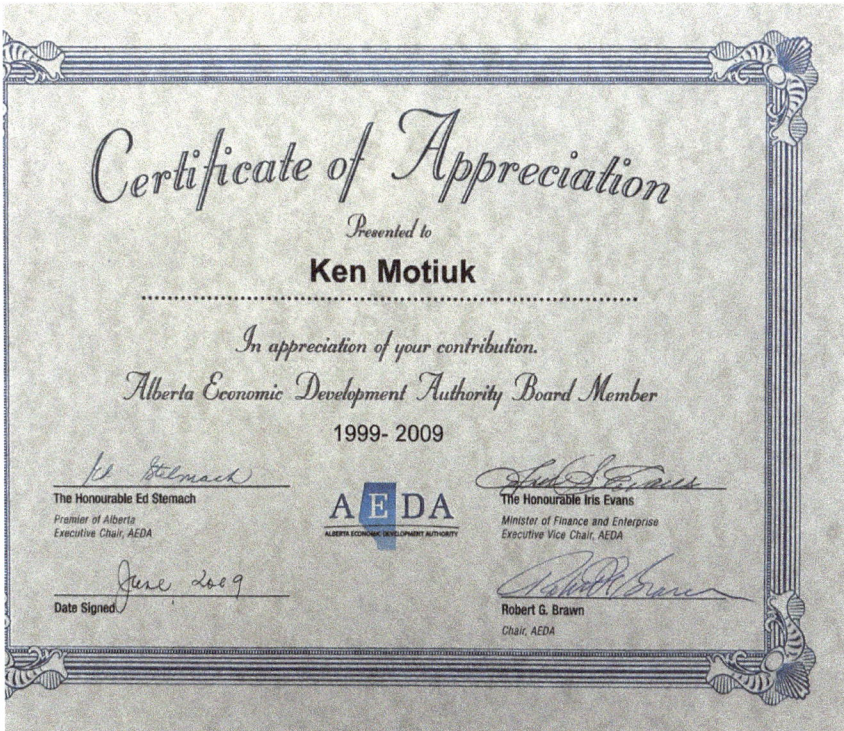

Certificate of Appreciation

Presented to

Ken Motiuk

In appreciation of your contribution.

Alberta Economic Development Authority Board Member

1999- 2009

The Honourable Ed Stelmach
Premier of Alberta
Executive Chair, AEDA

AEDA
ALBERTA ECONOMIC DEVELOPMENT AUTHORITY

The Honourable Iris Evans
Minister of Finance and Enterprise
Executive Vice Chair, AEDA

Date Signed June 2009

Robert G. Brawn
Chair, AEDA

CREDIT UNION DEPOSIT GUARANTEE CORPORATION

In the spring of 2003, I was asked by Bob Splane, a well-known member of Alberta's financial community, to serve on the board of the Credit Union Deposit Guarantee Corporation (CUDGC). CUDGC is the supervisory body established by the Alberta government to oversee the operation of the credit unions in the province. The Alberta government guarantees 100% of deposits in the credit union system and has a fiduciary interest in the financial health of these institutions.

The CUDGC board skill matrix was missing a member with knowledge of the agricultural industry as well as board-governance skills, so I accepted his offer. I served on the board of CUDGC from 2003 to 2013. While on the board, I served as chairman of the Audit Committee, and, for my last two years, chairman of the board. This experience broadened my knowledge of finance. During this time, I completed the Canadian Securities course and earned the designation of Certified Financial Advisor (CFA) in 2010.

My two years as chairman gave me insights into what responsibilities are like in a chairman's position. I stepped down from the CUDGC board in 2013.

A PEEK BEHIND THE CURTAIN—THE CWB

The Mulroney government in the 1980s knew the Canadian Wheat Board (CWB) should be changed, but they did not wish to directly address the issue. Many Mulroney supporters were Wheat Pool members and supporters of the CWB monopoly. Changing the Canadian Wheat Board Act would have created a great deal of controversy and the Liberals and NDP would have fought it vociferously in Parliament. Too much political capital would have to be expended by the Conservatives to change the Act.

Instead, the Mulroney government tried to change the way the CWB functioned by appointing commissioners they presumed would promote change from within. At this time, the CWB was run by five commissioners, all appointed by the federal government.

The CWB had a history of a powerful internal culture, reining in and converting new commissioners who came in with aspirations to change the way the CWB operated. In the 1970s, Forest Hetland, a successful and progressive market-oriented farmer from north-central Saskatchewan, was appointed commissioner by the Liberal government. Hetland was an early grower of rapeseed (now canola), a founding member of the Saskatchewan Rapeseed Growers Association, and an advocate for keeping rapeseed on the open market in the federal plebiscite in 1974. Once Hetland arrived at the CWB, his philosophy changed, and he became a strong CWB supporter and advocate for the single desk.

It was similar with Lorne Hehn, former president of UGG. UGG was a strong advocate of a deregulated, market-driven system, and Hehn spoke widely to promote this position on behalf of UGG. This policy approach to addressing system problems was established by Tom Crerar and Mac Runciman, previous presidents of UGG. Market-oriented farmers favoured UGG over the Wheat Pools, who conversely promoted collective marketing through the CWB.

In 1989 Hehn was appointed chief commissioner of the CWB by the Mulroney government. They hoped Hehn would bring the market-oriented solutions to system problems that he had advocated at UGG. Not to be! As soon as Hehn sat in the commissioner's chair, his views changed, and overnight he became an advocate of the single desk and defendant of regulations in the grain transportation system—the exact opposite of the views he held while at UGG.

So much for men of principle and the courage of conviction. The perks and prestige of this public office were simply too tempting for these political appointees to hold on to their earlier stance. Once in a commissioner's chair they both capriciously reversed their prior convictions and became strong supporters and advocates for the single desk. Opportunism trumped conviction.

Ken Beswick was appointed commissioner of the CWB by the Mulroney government in 1992. He was an advocate for change and maintained his free-market beliefs and principles while at the CWB and stood by them. However, he was shunned, ostracized, and shut out by the other four commissioners and senior staff, and kept out of the loop of day-to-day operations. He and his wife, Rosalyn, were not invited to the Christmas party held for commissioners and spouses. This behaviour attests to the character of the other four commissioners who attained their positions by political appointment rather than merit.

Adherents to the collective-marketing culture at the CWB would stop at nothing to discredit and demoralize those who did not share this culture. Beswick disgustedly resigned on his own accord in 1996, bitter to the end about the way he was treated at the CWB. He moved to Costa Rica and purchased a teak farm.

Throughout the Liberal governments of Jean Chrétien and Paul Martin from 1993 to 2006, the CWB retained its monopoly on exports of wheat and barley from Western Canada. This was largely due to the influence of Ralph Goodale, the only Liberal MP from Saskatchewan, who was a senior cabinet minister under both leaders. Ralph was a professional Liberal politician all his life, with little experience in the commercial world, yet he professed to know what was best for farmers and the grain industry. He was a stalwart supporter of the CWB.

By the 1990s, CWB support was waning in the farm community. Farmers did not like the lack of transparency and paternalistic attitude that surrounded CWB operations. They were becoming better educated and more secure, confident in their own ability to market grain. Modern information technology now allowed farmers from all parts of the Prairies to see from their farm desks what was happening daily in the world grain market. Oat marketing had been removed from CWB jurisdiction in 1989 and oat exports were increasing. As well, acres sown to canola, peas, and lentils, crops not marketed by the CWB, were increasing. These crops were marketed by the private grain trade, and farmers were becoming adept at marketing in this manner. The CWB influence over farmers from the days of King Wheat was no longer the order of the day. The farm community was prospering with crops that were marketed outside of CWB influence.

Millions of acres sown to CWB crops versus non-CWB crops

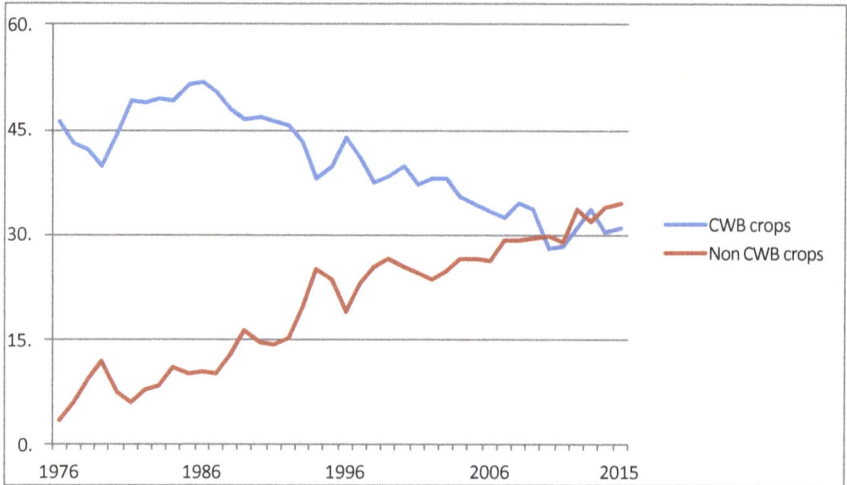

CWB crops include wheat and barley, and until 1988, oats

Non-CWB crops include canola, flaxseed, lentils, dry field peas, rye, chickpeas, mustard
and canary seed, and, from 1989 onwards, oats

Total seeded acres of main crops increased from 49.5 million in 1976 to 65.5 million in 2015. This came from a diminishing acreage of summer fallow.

In 1976, 93% of Prairie acreage was sown to crops marketed by the CWB. By 2015, this number was down to 47%. Farmers were voting with their seed drills. They were choosing crops they could market profitably on their own.

Pressure from the farm community to change the CWB mandate was mounting through the '90s. There were dueling studies completed on the CWB monopoly. The CWB hired professors Darryl Kraft and Ed Tyrchniewicz from the University of Manitoba, and Hartley Furtan from the University of Saskatchewan, to complete a study on the CWB. They concluded the monopoly accrued a benefit to the farm community. The Alberta government hired Ken Agra Consulting to complete a study on CWB barley marketing and the conclusion was the opposite. The Alberta government commissioned

Professors Colin Carter and Al Lyons to study the issue of the CWB monopoly on wheat and barley marketing and they concluded the benefits of the CWB, if any, were negligible. These reports were inconclusive, contradictory, and confusing.

The conclusions were dependent on the assumptions the authors made. Only the studies sanctioned by the CWB were allowed access to CWB documents. Other studies were not allowed this access, and then they were criticized by the CWB for not having all the information. The real question was: Should farmers abrogate their right to manage their own business in favour of maintaining the CWB, even when it is difficult to find measurable benefits?

In 1995, the Liberal government in Ottawa established the Western Grain Marketing Panel to study the CWB issue. The panel found that "weaknesses of the CWB system included bureaucratic inflexibility, inadequate servicing of specialized markets, lack of transparency, and the discouragement of value-added activities, particularly for smaller specialist processors."[50]

Recommendations for changes were made. These included more flexible options for sourcing grain from farmers, as well as more pricing options. The panel also recommended a change in the governance structure of the CWB and the returning of barley marketing to the open market. Goodale acted on some of the recommendations but refused to budge on the main recommendation and bedrock issue of removing barley from CWB control.

Totally frustrated with the ongoing public debate and inaction by the Chrétien government, some farmers started to take things into their own hands as they witnessed higher barley prices south of the border. These were markets they could not access without going through the CWB which offered them much lower prices. In 1996, Manitoba farmer Andy McMechan led the way with his own rebellious act of defiance. His action and personal sacrifice deserved support from other farmers. McMechan had been charged with failing to keep the peace after he hauled some wheat and barley

---

50 Miner, William M. *The Rise and Fall of the Canadian Wheat Board*. CAES Fellows. 2015–2. Page 20.

across the Canada–US border in Manitoba and refused to surrender his tractor as ordered by Canadian customs officials. McMechan was also charged with shipping grain into the United States without an export permit. His fines totaled $33,000 and he was ordered to pay the CWB $55,000 in compensation. McMechan also spent 155 days in jail and was treated like a hardened lifelong criminal while incarcerated."[51] All of this for simply selling his own grain!

A group of farmers became so disgruntled with these actions that they banded together into a group called Farmers for Justice in support of McMechan. After the McMechan escapade, members of this group blatantly carried sacks of wheat and barley across the US border at Coutts, Alberta, to make the public aware of how absurd the CWB's marketing restrictions were. Since the law is the law, border protection personnel were bound to arrest these farmers. The justice system had to prosecute them. Goodale and the Liberal government supported the prosecution. A few law-abiding Western farmers spent time in prison for simply taking their own grain across the border. This law did not apply in the other provinces of Canada—just the Prairies. Goodale, Chrétien, and the Liberal government stood by and watched. Hard to believe this was the twenty-first century!

---

51 Crawford, Russ, *Western Barley's Legacy: The History of the Western Barley Growers Association 1977–2022.* Agrinomics Publishing. 2022. Page 65.

On Market Freedom Day, August 1, 2012, Prime Minister Harper pardoned all thirteen farmers who were charged for their actions.

Left to right: Noel Hyslip (cut off), Ron Duffy, Bill Moore, Darren Winczura, Jim Chatenay, Gary Brandt, Rt. Hon. Stephen Harper, Rick Strankman, Jim Ness, Martin Hall, Ike Lanier, John Turcato, Mark Peterson, (missing) Rod Hanger.
Photo courtesy of Russ Crawford.

≈

In response to complaints from farmers that the CWB was too far removed and insensitive to the needs of the farm community, the government had earlier established the CWB Advisory Committee. In 1974, the make-up of the committee was changed, and its members became elected by farmers. Previously the members had been appointed by the government.

The committee was only advisory in nature and had no fiduciary responsibility. It consisted of elected farmers. As was to be expected when an election came up, the Wheat Pools' lobby would work hard to ensure that only farmers who supported CWB marketing were successful. This advisory committee, over time, lost its credibility and simply became a rubber stamp for the CWB.

Until 1997 the CWB was run by five commissioners who were appointed by the government and answered to Parliament. In 1997, in response to the Grain Marketing Panel, the government amended the CWB Act and changed the structure from that of a government agency to a shared-governance corporation. A fifteen-member board of directors would now govern the CWB as a shared-governance organization. Ten directors were to be elected by farmers, while the other five, including the CEO, would be appointed by the government. Other changes included the introduction of an optional contracting system to source grain from farmers, and various pricing options rather than only pooling. The Liberal government did not act on the main recommendation for barley marketing options.

Predictably the farmer-director elections turned into an ideological battle in the country over the CWB mandate. The voting process was skewed towards electing directors in favour of the traditional role of the CWB. For example, the vote was structured whereby eligible voters were both farmers and "interested parties" (landowners who rented out their land on a crop-share basis). Each would get one ballot on which to vote, no matter how much grain each person sold. What this meant was that an active, progressive farmer with a 3,000 acre farm would get one vote in an election. He was more likely to be inclined to welcome changes to the CWB mandate, as his livelihood depended on the success of his farm. However, if he rented land from three retired farmers (interested parties) on a crop-share basis, with 150 acres each, they would each get a vote as well. These older, retired farmers were likely more inclined to support the traditional role of the CWB, and they could outvote the actual farmer by 3:1 in an instance such as this. The voting process was skewed to preserve the status quo and could not reflect the interests of active and innovative farmers.

In the 1998/1999 crop year, the new CWB governance structure took effect. Ten elected farmers and five Liberal political appointees now governed the CWB. As expected, the farmer election returned a majority of leftist farmers and staunch CWB supporters. This group was not interested in reforming the CWB. In fact, they did the opposite and further infuriated farmers seeking reform by undertaking a campaign of advertising and self-promotion. This divided the farm community even further. The Liberal government in Ottawa no longer had to defend the CWB—the new board of directors did the bidding for them.

It is somewhat ironic that the CWB Act implies that farmers were not sufficiently intelligent to market their own crops. Yet the same Act allowed for these supposedly market-illiterate farmers to elect, from among their peers, representatives to guide and oversee the government-created agency responsible for the said function.

Greg Arason was appointed the first CEO of the new CWB. Greg, the former CEO of Manitoba Pool Elevators, had retired after the merger with Alberta Wheat Pool in 1998. He was a capable and experienced administrator who knew the industry well. He successfully bridged the operational transformation of the old CWB structure run by five appointed commissioners to the new structure of a CEO and board of directors.

≈

Meanwhile, the number of progressive, reform-minded farmers was growing steadily on the Prairies. Their frustration with the intransigence of the Liberal government and the CWB was increasing. In 1993, the Alberta Barley Commission and the Western Barley Growers Association, both farmer-funded groups, launched a challenge to CWB marketing under the Charter of Rights, claiming the CWB Act was an infringement upon their rights under the Charter. This case went to the Federal Court of Canada, which, in 1997, ruled the CWB did not infringe upon farmers' rights under the Charter.

Following came plebiscites where Western farmers voted on how they wanted their barley marketed. The Liberal government held a plebiscite, which allowed all farmers and interested parties, no matter whether they grew barley or not, to vote on how barley should be marketed. Over 60% of this group voted in favour of the CWB marketing of barley. The Alberta government conducted a plebiscite of only active barley farmers and found that over 60% wanted barley marketing removed from CWB jurisdiction. The results were clearly dependent on the wording of the question and the audience polled!

The Chrétien/Goodale era continued to be a frustrating time for the growing number of farmers who wished to see marketing opportunities other than the CWB. There had been a small degree of progress under the Mulroney Conservatives when they removed oats from CWB jurisdiction. The Mulroney Conservatives had also tried to move on barley in their eleventh hour with the Continental Barley Market but were unsuccessful when SWP challenged the move in the courts—and won. This Liberal government of Jean Chrétien and CWB Minister Ralph Goodale remained steadfast on the CWB's marketing monopoly.[52]

At the same time, the thorny issue of railcar allocation by the CWB remained even after the new Canada Transportation Act (CTA) was passed in 1995. With the elimination of the WGTA, the last remnants of the Crow Rate and Crow Subsidy were finally gone. Even though the entire system was moving to a more commercial system, the CWB still had control over railcar allocation. An industry-wide Car Allocation Group was formed to oversee the logistics of grain movement. The CWB did not wish to share this responsibility and wanted car allocation for CWB grain to be completely within its purview. The industry wanted to control the logistics of grain flowing through their assets and have the CWB take possession of CWB grain at port position. The CWB was totally uncompromising and would not relinquish any control over car allocation for their grain. This hybrid commercial/CWB system for car allocation was not working well. Grain companies were investing millions of dollars into new, large grain terminals,

---

52 The events around the issue of barley marketing and the Chrétien government are well explained by: Crawford, Russ. *Western Barley's Legacy – The History of the Western Barley Growers Association*.

and they wanted to be able to control the flow of product through these new facilities to their export terminals or other destinations.

In 1998, the federal government appointed retired Supreme Court Justice Willard Estay to conduct a review of grain transportation issues. Estey identified the CWB's centralized control over the system as a "bedrock" issue in the dysfunctionality of moving grain efficiently and competitively through the system. He recommended that the CWB's role in controlling grain-transportation logistics be terminated. Once again, the opponents of change, the CWB and left-wing farm groups resisted any changes to CWB authority. The Liberal government in Ottawa refused to challenge the CWB even after multiple conclusions that the problem in the transportation logistical morass lay in the hands of the CWB.

Next, the federal government appointed former Deputy Transport Minister Art Kroeger to head up an industry consultative group to implement Estey's recommendations. The battle between the CWB/left-wing farm groups on one side, and grain companies, railways, and progressive farm groups on the other, became so vociferous that Kroeger concluded "the historical divisions among Western stakeholders could not be bridged." CWB control over car allocation for their grain remained.

A stalemate hung over the debate of CWB reform in the first half decade of the twenty-first century. Any discussion regarding the power of the CWB led to an impasse. The number of reform-minded farmers continued to grow and they were becoming increasingly frustrated with the antics of the new CWB. The Chrétien Liberal government deflected all criticism of the CWB to the new board of directors. The new CWB shamelessly self-promoted with slogans and platitudes. Progressive farmers knew better. The monopoly powers of the CWB were hindering farmers' ability to manage their own businesses and seek the best prices for their wheat and barley. Crops marketed outside the CWB were growing in acreage and financial importance.

The new CWB board was functioning more like a left-wing farm group than a mandatory marketing agency. They promoted lost causes such as the doomed Port of Churchill and interceded into the debate on genetically modified organisms (GMO) crops in Canada. Funding for all of this was taken from the pool accounts of Western wheat and barley farmers who had no choice in the matter. Industry participants and Western farmers were waiting in hopes that the Chrétien Liberals would be ousted by the Harper Conservatives, who promised to change things.

Greg Arason retired in December 2002, and the Liberal Government appointed Adrian Measner as the new CEO. Measner was from Saskatchewan and had worked his entire career at the CWB. He was totally steeped in the historical Prairie lore of how the grain companies and railroads mistreated Prairie farmers "before they were saved by the CWB and single desk marketing." There was little hope of progress there.

On January 23, 2006, the Harper Conservative government was elected, albeit with a minority. Harper had learned all about the CWB earlier while on the Reform Party Agriculture Committee. Russ Crawford from Calgary worked in the grain industry and was a director of Preston Manning's constituency and a member of the Reform Party Agriculture Committee with Harper. The Reform Party vowed to change the mandatory marketing powers of the CWB at which so many farmers bristled. When Harper was involved with the National Citizens Coalition (NCC), he met Paul Orsak, a progressive and outspoken farmer from Binscarth, Manitoba. Orsak was a strong advocate of farmers' choice in marketing grain. Both Crawford and Orsak briefed Harper well on the shortcomings of the CWB. Harper did not like what he saw. He wanted to change the status quo and allow Western farmers the same freedom to market wheat and barley that farmers in other parts of Canada had.

With his minority government in 2006, Harper initiated change at the CWB. His first action was to remove the Liberal political appointees from the board and appoint members who were supportive of changes to the CWB mandate. Legislative changes were a problem. As the leader of a minority government, Harper knew they would have to make deals with the Liberals and NDP to change the CWB legislation. That was just not going to happen. And of course, the CWB would fight this initiative vehemently, using farmers' money from the pool accounts to do so.

In June of 2006, Liberal CWB board appointee Lynne Pearson's term expired. A vacancy was open. Pearson was from Saskatoon and had degrees in arts and Journalism.

In July of 2006, I received a phone call from Federal Minister Chuck Strahl's office in Ottawa offering me an opportunity to get back into the grain industry in Winnipeg. On September 15, 2006, I was appointed a director of the Canadian Wheat Board by the Stephen Harper Conservative government. I was the first of the Harper appointees to join the CWB at the September 2006 board meeting, assuming the position vacated by Pearson. This was finally a chance to get behind the scenes of the secretive CWB. The CWB was the last hurdle to a truly market-driven system of marketing, handling, and transporting grain in Western Canada.

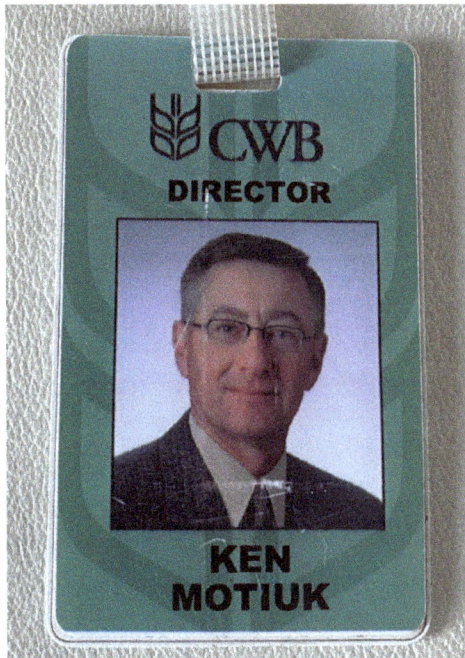

In October 2006, Ross Keith, a Regina lawyer and friend of Ralph Goodale's, was relieved of his directorship. In his place, Harper appointed Bruce Johnson, a capable veteran of the Western grain industry and most recently a senior executive with SWP. When SWP's finances hit the wall in the late '90s due to the capers of CEO Don Loewen, Bruce was Loewen's scapegoat when the financial mismanagement of Loewen became evident to the SWP board.

Also leaving in October 2006 was Bonnie Dupont, an oil industry executive from Calgary who was chairperson of the CWB Audit Committee. Of all the Liberal appointees, only Bonnie would have been an asset if she had stayed. She was an astute businesswoman and understood corporate governance. Prior to working with Enbridge, Dupont was human resources director of the Alberta Wheat Pool. But Harper wanted none of the Liberal appointees. In her place, Glen Findlay, a successful farmer from Shoal Lake, Manitoba, was appointed. Findley was formerly the minister of agriculture in the Manitoba Conservative government of Gary Filmon.

Remaining on the board was Adrian Measner, who was also the current CEO. The CWB Act stated that the CEO was to be appointed by the government upon discussion with the board of directors. Measner was a staunch supporter of the CWB single desk. He was not a visionary and had limited leadership skills. These shortcomings made him a weak CEO. But he knew CWB operations, and most importantly to the board of directors he was a passionate and unabashed supporter of the CWB. Harper removed Measner in December of 2006 and appointed former CWB CEO Greg Arason as interim CEO. Greg knew the CWB and the industry well, and his calm and capable personality was needed at the CWB at this time.

William Cheuk from Vancouver was the fifth remaining Liberal-appointed board member, having been appointed to the CWB board earlier in 2006 in the last days of the Paul Martin Liberals. He was not familiar with the Western grain industry or formal corporate governance but was a successful British Columbia businessman. Harper decided Cheuk would remain to complete his term.

Then came the ten elected directors. Two of them, Dwayne Anderson from Saskatchewan and James Chatenay from Alberta, were elected on a platform of reforming the CWB. The other eight were staunch CWB status quo supporters. Among CWB reformers they were disparagingly referred to as The Crazy Eight. Two of them, Rod Flaman and Ken Ritter, both from Saskatchewan, were originally elected on reform platforms. Once they sat around the CWB table they converted to CWB supporters. Just like Lorne Hehn and Forrest Hetland before them, they drank the Kool-Aid!

Ken Ritter was a lawyer and part-time farmer from the Kindersley area, and politically a Conservative. His political views did not coincide with his being a strong CWB supporter. Ken became the first chairman of the board under the new governance structure. Tempering his CWB reform views likely helped him to become, and remain, chairman. Larry Hill was a successful and progressive farmer from the Swift Current area and was without a doubt the most capable and broadminded of the group that supported the CWB mandate.

The remaining six elected farmer-directors were hardcore CWB supporters, most of them members of the hard-left farm organization, the National Farmers Union (NFU).

In an election of farmer members in the fall of 2006, Dwayne Anderson's position fell to Kyle Korneychuk, another passionate CWB supporter. In Alberta, Henry Vos defeated Art Macklin in the election in the Peace River district. Henry supported CWB reform. Among the ten farmer-elected members, eight supported the status quo and two supported changes. On this fifteen-member board, CWB reform was not going to originate at the board level. The fight with the Harper government was on. Eight of the board members were there to support and promote the status quo at any cost. This was not corporate governance! Directors were elected in the country on a platform based on where they stood on the monopoly rather than their corporate-governance ability and business skill set. Ritter remained chairman after the 2006 farmer-director election.

CWB Board of Directors, September 2006

Left to right: Art Macklin (AB farmer), Jim Chatenay (AB farmer), Bill Toews (MB farmer), Ken Ritter (SK farmer), Bill Nicholson (SK farmer), Rod Flaman (SK farmer), William Cheuk, Allen Oberg (AB farmer)

Left to right: Ian McCreary (SK farmer), Dwayne Anderson (SK farmer), Ross Keith
Adrian Measner (CEO), Larry Hill (SK farmer), Bonnie Dupont, Ken Motiuk

The first CWB board meeting I attended was in Winnipeg on September 27 and 28, 2006. I was the first and only Harper appointee at this meeting. I was asked by Chairman Ritter to serve on the Audit Committee and the Governance and Management Resources Committee. Going into my first board meeting was somewhat unnerving, to say the least. I was walking into a situation where my beliefs were directly opposed to those of the majority in the room.

There was a rigid seating structure around the table based upon seniority of tenure. That put me at the furthest end of the table from the chairman. Many of the board members lacked social decorum, business acumen, and financial knowledge. This meeting was a real gong show! The CWB was run by a group of financial Neanderthals. Their main interest was mindlessly supporting and promoting the CWB without question.

Despite all this, finding myself in this boardroom was initially quite intimidating. It prompted a sense of history: Many historic events and decisions that affected Western Canadian agriculture had occurred in this room. It was where the Soviets met to purchase wheat from Canada in 1972. For many years, five powerful commissioners met in this room and made decisions that affected the entire Prairie grain economy. Now I had the chance to be one of the players. I was in the lion's den of Canadian wheat marketing.

Bruce Johnson joined us at the December 2006 meeting and our group of reformers now included Jim Chatenay, Henry Vos, Bruce Johnson, and me. Chatenay had been the most vocal member of the board in challenging the status quo. At times, when Jim would continually question the actions of the board, the chairman would ask him to leave the board table and sit in the outside lobby for a while—the equivalent of a kindergarten timeout. The atmosphere the left-dominated board maintained was not conducive to a full and complete discussion on any issue they did not agree with. It was a strict Stalinist adherence to the doctrine or be shamed, shunned, and banished!

At this point, the level of congeniality between the CWB board and the Harper government deteriorated to a series of taunts, challenges, confrontations, and public battles. The board was dominated by the eight left-wing elected farmers. It was not possible to sway any of these board members with logic, so their decisions were the face of the CWB. The Act stipulated that the CEO was to be appointed by the federal government and the CWB was to pay his salary. In protest of the removal of Measner and the appointment of Arason as CEO, the board refused to pay Arason. Minister Strahl had to issue a directive to the CWB to pay Arason.

The Harper government issued another directive to allow barley to be marketed outside the CWB. This was essentially a jurisdictional dispute. The CWB board refused to comply and challenged this directive in court. The court upheld the challenge, as it stated that the Act must be changed to allow for barley to be marketed outside the CWB. Late in 2008, the government introduced legislation to change the Act and remove barley marketing from CWB jurisdiction. Unfortunately, it died on the Order Paper when the October 2008 election was called.

The government issued another directive that the CWB stop spending farmers' money on self-promotion of the single desk. The CWB challenged this in court as well and once again won. The CWB board engaged in all types of obstructionist tactics to frustrate the Harper government. The stalemate continued, and the relationship between the federal government and the CWB deteriorated even further.

At one point, when I met Minister Strahl in an airport where our travels crossed, he laughingly apologized for giving me the "worst government appointment ever."

During this time as a CWB director, I pursued my professional development by enrolling in the Directors' College program offered through Dalhousie University. This was a five-module program established to develop and enhance the skills of directors of companies. I completed this program and earned my Charter Director (C. Dir) designation early in 2009 after passing the exam.

When Greg Arason took on the position of CEO in late 2006, it was meant to be a transitional measure, and by late 2007, Greg wanted to leave. Early in 2008, the CWB board undertook a search for a new CEO. The leftist farmer group on the board convinced former CEO Adrian Measner to apply for the job. They wanted to put Measner forth as the board selection for CEO, knowing it would further irritate the Harper government. Meanwhile, Ian White, a capable and experienced grain executive from Australia, was recruited by Bruce Johnson and his name was put forth. Johnson had come to know White in the 1980s when White worked with the Australian company Elders Grain in the Canadian grain industry.

Measner's main (and some would suggest only) attribute was an unwavering belief in preservation of the single-desk CWB.

The selection for a CEO came down to Measner and White. It was a secret ballot, and Arason did not vote. It looked like the vote would turn out 8–6 in favour of Measner, as most votes went.

However, to everyone's surprise, the vote came in 8–6 for White, who was by far the best qualified candidate. Interestingly, two of the CWB supporters jumped ship and voted for the best business candidate rather than for the strongest CWB supporter. White was appointed CEO in March of 2008.

Other changes on the board occurred in 2008. Ken Ritter resigned as chairman and Larry Hill took over chairmanship. During his last while on the board, Ritter became a supporter of reform once again. Ritter retired and did not run in the 2008 director election.

In 2008, the term of William Cheuk, last of the Liberal appointees, expired. The Harper government reached out to several Conservative supporters including Paul Orsak and myself to recommend a suitable nominee. Orsak and I had worked together on the Wheat Growers, and we stayed in touch when Paul went back to his family farm in Manitoba, and I continued to farm in Alberta. Because the board did not have an accountant as a member, we suggested David Carefoot, a former financial executive with UGG and Agricore United. Carefoot was an accountant who understood the Canadian grain industry well. The Harper government wanted someone who was broadminded and had strong business credentials so the appointment would be above reproach by political critics. Paul Orsak took this to his connections in the PMO and Carefoot was appointed to the board in May 2008.

On October 14, 2008, the Harper Conservatives were elected once again with a minority government and Gerry Ritz was appointed minister of agriculture with responsibility for the CWB. Ritz was from Saskatchewan and a strong proponent of CWB reform. But again, the government was hampered by their minority. It would be difficult to pass legislation in Parliament to change the CWB Act, as the opposition Liberals and NDP would never allow it. So, the stalemate between the CWB and the government continued, along with the taunting, haughty, and belligerent attitude of the hard-left directors. Their fondest wish, a defeat of the Harper Conservatives, had not occurred.

CWB Board of Directors 2008–2009

Left to right – Bill Toews (MB farmer), Cam Goff (SK farmer), Kyle Korneychuk (SK farmer)

Left to right – Jeff Nielsen (AB Farmer), Bruce Johnson (appointed), Glenn Findlay (appointed), Henry Vos (AB farmer)

Left to right – Bill Woods (SK farmer), Ian White (CEO), David Carefoot (appointed),
Bill Nicholson (SK farmer)

Left to right – Larry Hill, (Chairman and SK Farmer), Rod Flaman (SK Farmer), Ken Motiuk (appointed)

Chairman Hill and CEO White worked well together and provided a calming influence on the board, diplomatically attempting to bridge and appease the two factions. Bruce Johnson, Glenn Findlay, and I were reappointed for another three-year term in 2009. Nothing much happened in those couple of years before the 2011 federal election. Hill left the position of chair and did not run again in the director election of 2010. Stewart Wells, president of the NFU, replaced Hill in the director election, so this was another hard turn to the left. The chair was taken over by Allen Oberg, a farmer from Alberta with weak board credentials, to whom even mediocrity was a challenge. There was to be no reaching out to the government for any kind of peace treaty!

The reform members on the board were waiting for Harper to win a majority, while the CWB stalwarts were waiting for the Liberals to win the election. More and more Western farmers viewed the CWB as an anachronism.

THE PRAIRIE GRAIN ECONOMY BOOMS

Grain prices around the world had been in a malaise since the early 1980s. After the Soviet Grain Robbery and the Arab Oil Embargo in the 1970s, commodity and oil prices skyrocketed, and inflation went up to 14%. Interest rates increased dramatically to try to cool down world economies. By 1981, interest rates were around 20%. Farmland prices increased about tenfold between 1971 and 1981. Grain prices settled into a new trading range that was higher than before the boom but lower than the highs of the 1970s. Increased production soon resulted in a surplus of world grain supply and then very soft prices. All this, combined with the high interest rates, resulted in a farm debt disaster in the rural community.[53] Farmland prices dropped.

World oil prices fell as well, and to make it worse on Western Canada, the Trudeau Liberal government imposed the National Energy Program on the West. Now both agriculture and the energy industry had the rug pulled out from under them. Residential and farmland prices which peaked about 1981/1982 started to fall. It took over twenty years until about 2005 for housing and farmland prices to reach and exceed the former peak of 1982.

As grain prices fell, competition between grain-exporting countries intensified. The EU and the US resorted to using export subsidies to increase their respective market shares. Stocks of grain

---

53 See the section on the Farm Debt Review Board (Page 133)

went up and prices went down. The federal government would not introduce a similar program in Canada, so Western grain farmers had to tough it out, competing against subsidized exports from the EU and the US.

This cycle did not end until the George W. Bush administration took over in the US in 2001. That Republican administration decided they were going to terminate direct farm production and export subsidies. Instead, they would use the money to fund growth of the ethanol industry using corn as a feedstock. This was more appealing politically as it eliminated farm subsidies and redirected the funds into a renewable fuel–production program. This favourably changed the economics of agriculture. Ethanol production used up the surplus supply of corn, and grain and oilseed prices started to increase.

By 2008, grain prices rallied considerably and we were into a boom cycle in Prairie grain and oilseed production. Canola prices reached as high as $14.00/bushel. Profitability of other crops followed. Farming became exciting and rewarding again. This boom lasted for close to a decade. Farmland prices rose rapidly as well, with the price of farmland roughly quadrupling during this period.

At our farm in Mundare, we were on a new tier of growth. Consolidation of smaller farms was occurring rapidly. Our daughter Carlee and her husband Justin Leliuk joined us on the farm. We had continually been growing the farm, and now we had a new impetus to keep doing so. Our farm succession plan was evolving.

We were seeding about 6,000 acres in 2008 and actively looking to expand further. We used the profits from good commodity prices to upgrade our farm equipment and prepare ourselves for further farm expansion.

THE FINAL FIVE

On May 3, 2011, the Harper Conservatives won a majority government. Gerry Ritz remained minister of agriculture with responsibility for the Canadian Wheat Board. With a majority government, the path was now open to make changes to the CWB Act.

In October 2011, the government introduced the Marketing Freedom for Grain Farmers Act. The CWB monopoly on Western wheat and barley marketing was to end on August 1, 2012. CWB authority over railcar allocation and system throughput on wheat and barley would also finally end.

The ten elected farmers were removed from the governance structure and the five remaining appointed directors, Johnson, White, Findlay, Carefoot, and Motiuk, became the new board of directors. Johnson took over as chair, and White continued as CEO.

We were instructed by the government to wind down operations of the CWB within five years. The Act became law on Dec 15, 2011. In preparation for the upcoming change, the first teleconference meeting of our new board was held on November 4, 2011. On January 10, 2012, we met with Minister Ritz in Winnipeg.

Five Member CWB Board, 2012

Left to right: Ian White (CEO), David Carefoot, Bruce Johnson (Chairman), Glenn Findlay, Ken Motiuk

And so, the transition of the CWB from a government monopoly to a private entity began. Operations were scaled down and the organization continued to export grain, purchasing in competition with the private trade. Since the CWB did not own any country facilities, arrangements were made with existing companies to handle grain for the CWB. Opportunities to purchase country facilities were pursued.

The CWB's staff was downsized, as fewer personnel were needed to operate at the new level of activity. A core group of four senior CWB executives was retained to oversee the transition. Ian White continued as CEO, Ward Weisensel oversaw operations, Brita Chell continued as CFO, and Dana Spiring continued as legal counsel with responsibility for corporate development.

The CWB began to search for opportunities to purchase facilities to grow the new entity. None of the mainline companies wished to part with facilities, so for acquisitions the CWB looked to smaller players as well as the new farmer-owned facilities that had sprung up around the Prairies. Sites were researched for the construction of new terminal grain elevators.

The CWB purchased Mission Terminals in Thunder Bay along with its small grain-handling assets on the Prairies. They also purchased Les Elevators des Trois-Rivieres and Services Maritimes Laviolette in Quebec. The first large collection facility they purchased on the Prairies was Prairie West Terminal, located between Dodsland and Plenty, SK. The purchase included a farm input–supply business, as well. In 2014, two sites were chosen to construct modern new facilities: Colonsay, Saskatchewan, and Portage La Prairie, Manitoba. There were plans to prepare a site at Pasqua, Saskatchewan, as well.

Throughout this transition period, the board of five members, along with the four top executives, met several times with Minister Ritz to provide updates on our progress. Our ministerial briefings usually took place in Winnipeg, with Minister Ritz often accompanied by his assistant, Devin Dreeshen.[54]

We also started looking to what the long-term future of the new CWB would look like. We had been told by Minister Ritz that the government would like to see that any new mergers or acquisitions we engaged in would ultimately result in more competition in the Prairie grain industry. We studied the grain industry from an international perspective to see if we could find any suitors for a merger. A new stand-alone CWB would have difficulty competing in the long term against the larger, more established companies currently operating on the Prairies.

---

54 Dreeshen went on to become a cabinet minister in the UCP government in Alberta.

After much study, analysis, discussion, and negotiation, we decided to partner with a new company, G3 Global Grain Group, which was a joint venture between Bunge and the Saudi Agriculture and Livestock Investment Company (SALIC). Not only was this an entirely new company that had plans to build facilities on the Prairies, but along with it came a modern new export grain terminal in Vancouver that Bunge proposed to build. This would become the first new grain terminal built in Vancouver in decades.

On July 31, 2015, the newly merged company was announced and rebranded as G3 Canada. The old CWB no longer existed.

Arrangements also had to be made for farmer ownership of the component of the company that came from the old CWB. The Farmers Equity Trust was established, whereby this ownership would go to farmers who did business with G3.

G3 went on to build more country facilities and the new Vancouver terminal was completed. Farmers Equity Trust continues to manage the farmers' share of investment in the prospering new company.

July 21, 2015, was the last meeting of the board of the CWB. As we adjourned, my parting words were, "Now I can go back to the farm and make money growing wheat." And that's exactly what I did!

NO, CHICKEN LITTLE, THE SKY DID NOT FALL

For decades we were told by the agri-political left that if the CWB were to be disbanded it would lead to the financial ruin of the Prairie grain economy. For a long time, the propaganda machines of the CWB/Pools/NFU preyed on the fears and insecurities of Prairie farmers and managed to convince most farmers to believe this myth. In fact, quite the opposite transpired.

August 1, 2022, was the ten-year anniversary of Marketing Freedom Day, whereby farmers were able to sell their own wheat and barley to any company for the first time in over seventy-five years. Presently the industry is flourishing, and grain production and farm income have increased. Billions of dollars have been invested in system improvements by the railroads and grain companies. New grain-receiving terminals in the country have been built by various companies. G3/Bunge built a new terminal with an unloading loop rail track on the north shore of the very crowded Burrard Inlet. Fraser Grain, a partnership of Parrish & Heimbecker and Grains Connect, built a new grain-export facility near the outlet of the Fraser River, the first such facility built outside of Burrard Inlet.

The export of grain, oilseeds, pulse crops, and other field crops has increased. Just as when oats were taken off the CWB, exports for barley started increasing after the CWB monopoly over exports was removed.

There have been large investments in value-added facilities on the Prairies. The deregulation of the Prairie grain industry has been an astounding success.

Courtesy of G3 Canada Ltd.
The G3 Terminal above is at Maidstone, Saskatchewan. It includes a loop track for a 130-car unit train. Cars are loaded in the country in a continuous loop and then pulled to the Burrard Inlet as a unit train and unloaded in that same continuous loop at the Bunge Terminal.

Our trucks hauling wheat to the Richardson Pioneer facility at Lamont, Alberta, a modern country grain terminal.

Farmers now haul their grain to large country grain terminals with storage capacities of up to 45,000 tonnes. These facilities utilize a loop-train concept for loading whereby dedicated grain trains up to 150 cars long come in to pick up the same grain in a large loop with the train never having to be decoupled into smaller segments. This continuous train can then go straight to Vancouver and be unloaded at the new G3 facility, which utilizes a similar loop to unload the railcars without detachment.

> G3 has recently set a record for transporting Western grain. One 150-car loop train was able to make two complete trips from Alberta to Vancouver within one week.

In the mid-1970s the size of the boxcar used to haul grain was about sixty tonnes, the same size as the boxcars used fifty years earlier in the 1920s. An eighty-car train would pull 4,800 tonnes to tidewater after collecting the widely dispersed cars from numerous country elevators. By the mid-1990s the average size of a hopper-car was 90 tonnes, and a 100-car train would pull 9,000 tonnes, almost twice that of twenty years prior. By 2023, new hopper cars held 110 tonnes, and a 140-car unit train, loading at one country facility and unloading at one tidewater terminal, can pull 15,400 tonnes. Once compensated for hauling grain, the railroads responded with massive investments in new rolling stock and technologies.

Forty years ago, under the regulated system, the average turnaround time for one boxcar was more than twenty-two days. Components of the grain-handling-and-transportation system are now functioning much more efficiently than when they were controlled by the CWB.

In October of 2022, both CN and CP Rail set both individual and system records in terms of grain shipped to port position. In the fourth quarter of 2022, the Port of Vancouver set a record for grain loaded on vessels.

A summary of the successful transformation of the system is well documented in a paper by Quorum Corporation.[55]

A summary of some of the salient findings by Hemmes:

1. Average annual production of the main crops on the Prairies went from 50 million metric tonnes (MMT) in the 1990s to a current level of over 70 MMT.

2. Exports of these same crops went from 25 MMT to over 40 MMT in the same period.

3. There has been a shift in export shipments from Thunder Bay to the West Coast.

4. With elimination of the mandate of the CWB, "Management of the logistics of grain movement became more efficient with the grain companies now having complete control over the management of their assets in the country and at port. Ultimately, new investment in the system began to flow at an unprecedented level with new elevators, increased storage, and improvements to the efficiency of port terminals. It also brought new players into the industry - both foreign and domestic."

So, what does the farm community in Western Canada think about all this, ten years after we achieved market freedom? *Real Agristudies* conducted a survey of 480 prairie farmers in August of 2022 to assess the sentiment among farmers as to the success of the change. Of those who responded, 77% indicated they felt the change turned out better than expected. Most respondents also felt Canada's global wheat and barley reputation has increased since 2012.

Any way you look at it, the deregulatory changes that occurred in the Prairie grain industry in the last forty years have been an outstanding success. There is now surplus capacity in the system to be able to handle spikes in exports when they occur. Allowing

---

55 Hemmes, Mark A. 25 Years of the Grain Handling and Transportation System (1995-2020): A Time of Great Change. Quorum Corporation. 2021.

market forces to work has solved the problems created by misguided regulations and bureaucratic incompetence.

The most recent Transportation Modernization Act creates an environment much more favourable for railway investment than that of the past.

Historically, many rail shippers were distrustful of rates set by railways since they were essentially monopolies along their lines. A shipper on a rail line becomes captive to that railway. As a result, the federal government closely monitors any rate changes railways make to ensure that no unreasonable rates are set. This oversight is true for grain shipments as well. It is called the Maximum Revenue Entitlement (MRE), and both CN and CP Rail are only allowed to make a certain level of profit from grain movement annually as set in the MRE. The metrics of this are monitored by Quorum Corporation.

Agricultural profitability manifests itself in property values. The following table, using Farm Credit Corporation (FCC) data, depicts farmland values in northern Alberta from 1985 to 2021. There has been a substantial increase from about $400/acre in 1985 to $4,400/acre in 2023. Most of the increase has occurred since 2005, somewhat coincidently as farmers were achieving more freedom in managing their grain sales.

Farmland value per arable acre in northern Alberta

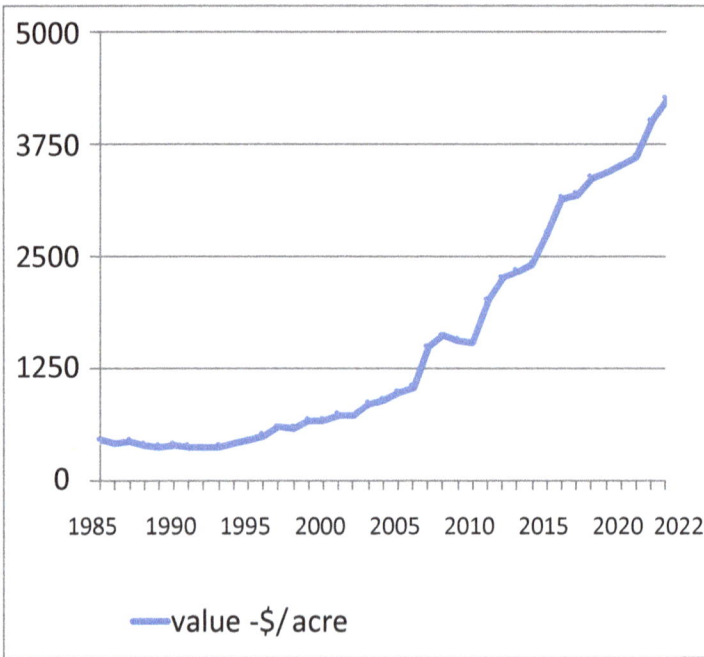

Source – Farm Credit Corporation

In my 1979 letter to Don Mazankowski (Appendix #2 to this paper) I postulated that until two changes were made to the regulatory environment in grain transportation and marketing, the industry would not be able to prosper. These two changes were: eliminating the old Crow Rate and paying the railroads to transport grain; and eliminating the single-desk grain-marketing mandate of the CWB. These are now both gone, and the industry is indeed prospering. The debacle of collective marketing and the Dark Ages of the CWB and the regulated system are finally past us.

Thank you to both Prime Minister Stephen Harper and Agriculture Minister Gerry Ritz for enabling this long-simmering issue that hampered the prosperity of Western Canadian agriculture to finally be resolved.

Stephen Harper and Gerry Ritz addressing a group
of farmers on Marketing Freedom Day

Market Freedom Day, August 1, 2012, Art Walde's farm Kindersley, Saskatchewan.
Courtesy Russ Crawford

MEANWHILE, BACK AT THE FARM

Our farm operation is now a multigenerational business. Our daughter Carlee and her husband Justin Leliuk came back to the farm in 2008. In 2011, our family agreed on a formal succession plan to migrate management responsibility from myself and Wendy, to Carlee and Justin. A plan for transferring our farm assets to the next generation was developed, allowing for stable continuity of the farm as well as providing for our non-farming daughters. Everyone agreed to the succession plan, and we are still guided by it. Since then, Carlee and Justin have capably taken over and substantially grown the farm. They are reaping the benefits of their own efforts as I retreat from responsibilities.

The Motiuk farm has grown from 300 seeded acres in the 1960s to a current level of 13,000 acres. I assist with bookwork, finances, and planning. I help with spring planting and autumn harvest. We spend our winters at our Arizona home. I am semi-retired and living in the best of both worlds—participating in our growing family farm with the next two generations and taking time to enjoy travel and retirement activities in the off-season.

Changes to farming practices have made the land more productive and the farm more profitable. Crop yields are up. In 2023, our farm experienced the best crop ever, with record yields in all crops. The innovative practices we adopted over the years that were ahead of most of the farm community served us well, as we have a competitive advantage over other farmers in the area.

The changes to the marketing system allow us to use our marketing skills to the best advantage of our farm. Under the CWB system, we were merely producers of wheat. With the CWB gone, we are now managers of our complete farm business. Good marketing is a profit centre on our farm.

Growth of Motiuk family farm: Acres sown to annual crops

1977 - 2023

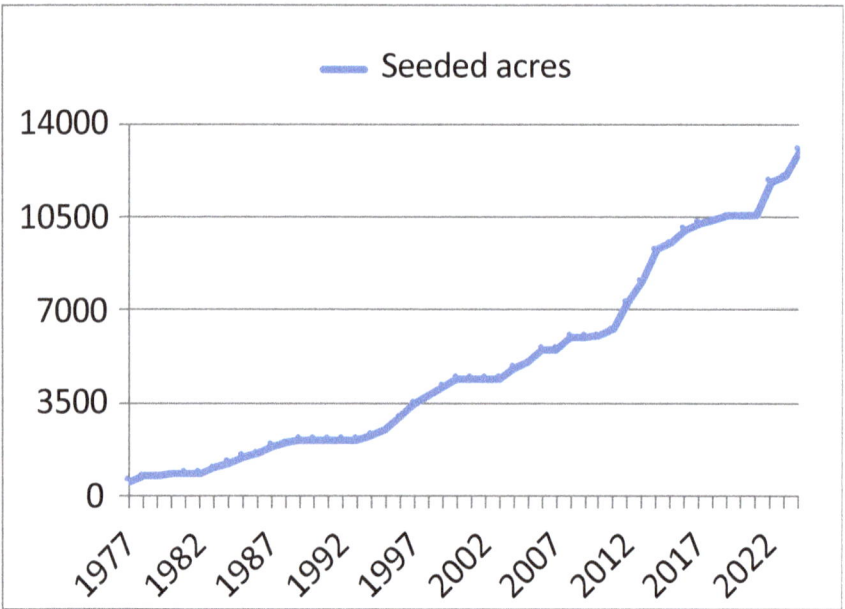

Astute farm managers can now plan their grain deliveries to match cash flow and farm storage restrictions. As price rallies occur in months prior to harvest, contracts for pricing and delivery can be made right off the combine to meet farm cash flow and logistical needs. Once a contract is made with a grain company, space will be saved at the elevator so that the farmer will be able to deliver the grain at the allotted time. The grain companies can plan their grain throughput using this contracting process, and when they have enough grain for a full train, rail shipment can be completed. The elevator companies can manage their storage space, and the railroads can manage their train bookings.

None of this was possible under the CWB's quota system. It was impossible for a farmer to be able to complete a budget or a business plan when he did not know when he would be able to deliver his grain, or what the final price would be, until sixteen months after harvest when the CWB final payment was made. In the days of the CWB, every farmer would be allowed a 3 bushel quota, until all farmers across the Prairies filled their quota, regardless of individual cash and logistical needs. Then another quota would open in a similar fashion.

The elevator companies were compelled to take the grain according to whatever the quota was. The CWB would then allocate individual grain cars to each company based upon a formula calculated by past handlings. This method did not serve to optimize efficient and effective grain movement. Now, those who own the assets can manage their use.

We farm a large block of land about twenty-five miles from our main farm. It happens to be located near two large grain terminals. When we harvest there, we will contract the amount of grain we expect to harvest for delivery directly from the field to the elevator. Since we have a contract, space is allotted for our delivery. It saves us having to haul our grain twenty-five miles back to our main bin yard and then haul it back to market later in the winter. This type of practical and efficient logistics management would not have been allowed under the CWB quota system.

Country elevator managers know how much grain they have contracted monthly, and they allocate the space for it to be delivered. Then they book trains far ahead to ship this contracted grain to market. The owners of the facilities manage the throughput.

Whenever I faced challenges and obstacles over the years, I always thought back to a quote from management consultant Jim Collins in his book *From Good to Great*: "What separates people [...] is not the presence or absence of difficulty, but how they deal with the inevitable difficulties in life."[56]

In the early 2020s, Prairie grain production was in another period of boom. Worldwide supply chain disruptions due to the COVID pandemic resulted in commodity prices increasing. The Russian invasion of Ukraine in February of 2022 further shocked world food production and supply. In early 2022 farm gate canola prices hit $25.00/bushel and wheat $15.00/bushel. Prices have now receded, but the world is becoming more appreciative of food security and secure production capabilities. We now have a market-driven grain-marketing-and-transportation system that can handle increased demands placed upon it.

---

56   Collins, Jim. *From Good to Great*. HarperCollins Publishers. 2001. pg.85-86.

The way is paved for the fifth generation of Motiuk descendants to continue farming the
land homesteaded by my grandfather in 1899.

Kaden, Taya, and Jace Leliuk in a field of blooming canola.

Fifth generation on the Motiuk farm.

Motiuk Family Farmstead - 2019

SUMMARY AND FINAL THOUGHTS

WORLD GRAIN TRADE

In the early 1900s and through World War I, Canada was a major player in the worldwide wheat-export market. At this time, wheat was the main crop traded internationally. World trade dynamics in wheat started to shift after the Russian Revolution of 1917. Russia ceased to be an exporter of wheat. In these early years through to World War II, Canada supplied nearly 40% of the worldwide wheat-export market.

Through the 1920s, Europe restored their wheat stocks, and the US increased their wheat production. Two smaller players, Australia and Argentina, became exporters of wheat as well. These four entities, along with Canada, dominated wheat trade in the world right through to the 1980s. By the 1980s, the largest wheat exporters were the US and the European Union (EU), and then Canada, followed by Australia and Argentina. Canada's share of the world market fell to about 22%. Currently, Canada supplies about 13% of wheat worldwide (last seven years average). Accordingly, Canada's influence on the worldwide wheat market diminished substantially through the twentieth century. Canada is now the sixth-largest exporter of wheat in the world.

Dynamics in world trade of grain changed after the pivotal shift triggered by the Soviet grain purchases in the early 1970s. In

Canada, canola acreage increased, and we became a world leader in the canola trade. Canada is now the world's largest exporter of canola, and we have a strong canola-processing industry. Oilseeds became a significant component of world trade, and it became more appropriate to start referring to world trade in grains and oilseeds rather than just grain. Soybean acreage in the US also increased. Brazil went from having very low soybean production in the 1970s to being the largest producer and exporter of soybeans in the world today.

A similar situation occurred with palm oil in Malaysia. Palm oil production was negligible in Malaysia in the 1970s and with the growth of this industry Malaysia is now the largest producer and exporter of palm oil in the world.

A most spectacular change occurred in corn production in the USA. Through hybridization of seed and the application of GMO technology, average corn yields increased from 40 bushels per acre in 1950 to 177 bushels/acre in 2023. As well, the US had been the largest producer and exporter of soybeans until the early 2000s, when Brazilian production and exports of soybeans surpassed that of the US.

New production technologies were so successful in the 1980s and 1990s that the world had a significant surplus of grain, which resulted in low prices for farmers worldwide. The US and EU began to pay their farmers large production subsidies for wheat and corn. This all changed in the early 2000s when the George W. Bush administration in the US redirected farm subsidies. Rather than going to farmers directly, these funds subsidized the production of ethanol from corn. Corn is a renewable resource that now provides some of our energy needs. Currently 38% of American corn production goes to ethanol production. This uses up the large supply of corn, and market prices are now high enough that farm-production subsidies are no longer necessary.

Similarly, with oilseeds such as canola in Canada and soybeans in the US, as production of these crops expands, increasing amounts are used in the production of biodiesel, another renewable source of energy.

World dynamics in production and trade of grains and oilseeds changed again after the breakup of the Soviet Union in 1991. With the failure of the disastrous Soviet experiment of collective farms, these farms were disbanded, and the land was returned to former owners and employees of the collective farms. Oligarchs and private European agricultural interests took over these tracts of land and improved them into large, modern, productive farms. By 2020, Russia, Kazakhstan, and Ukraine together became the largest exporters of wheat in the world. These three together are referred to as the Black Sea production area. These dynamics changed again after Russia invaded Ukraine in 2022, and now Ukrainian production and export of grains and oilseeds has been hampered by the military disruptions of the Russians.

Below is a table of worldwide wheat production by country or region. Followed by Europe, China and India are the largest wheat producers. However, because of their large populations, these two countries are still importers of wheat. The largest exporters are countries with large land bases and low populations. Canada only produces 3% of the world's wheat.

Ten-year average (million metric tonnes) of worldwide
wheat production by country 2013–2022.

Europe		139.04
China		131.06
India		97.92
Russia		67.34
USA		53.93
Middle East		33.36
Canada		30.73
Ukraine		25.65
Australia		24.72
North Africa		18.3
Argentina		16.1
Kazakhstan		13.18

CANADIAN WHEAT BOARD

The CWB was an appropriate policy instrument for its original purpose, that of overseeing the price, supply, and distribution of grain in war-time periods. It was not successful during normal times. It also bears repetition that the CWB was not a farm subsidy. Though controlled by Ottawa, farmers paid all the expenses of the CWB.

> Like other governmental and quasi-governmental bodies, over time the CWB fell victim to a bureaucratic mindset, jealously guarding its power and authority, assuming an attitude of superiority, and failing to evolve with changing times. It eventually became an impediment to the farmers it was supposed to serve.

Many of the policies of the CWB were detrimental to the growth of the Prairie grain industry. Delivery quotas based on summer fallow acres annually resulted in millions of acres not being sown to a crop and left idle on the prairies. Much of this land could have been seeded to crops to provide revenue to farmers and the Western economy. Having the unprotected soil exposed to the elements caused soil degradation due to wind and water erosion.

In the very dry southern Prairies, often known as the short-grass country or brown-soil zone, there was a greater need to summer fallow as two years of precipitation was often required to store sufficient moisture to grow one crop. However, the Prairies are not all dry, short-grass country. Further northward in the dark brown–soil zones, and then even further north to the moist, fertile, black-and-grey wooded soils, fallowing was not necessary for moisture retention. Rainfall is adequate and crop yields are higher. The one-size-fits-all quota system the CWB employed, heedless of agronomic realities, resulted in tens of thousands of acres of fertile cropland fallowed in the more northern parts of the Prairies so that farmers there could get enough quota to sell the extra production from their more fertile land.

In the late 1960s and early 1970s, the CWB withheld export sales of wheat in the misguided belief that if Canada held back wheat sales, the world price would go up. By this time, Canada was not a large enough player in the world market to be able to influence price. During this time, Prairie farmers were desperate for cash flow due to the low delivery quotas the CWB offered. The politically appointed commissioners at the CWB were oblivious to the tremendous financial stress on Prairie farms. This is a prime example of the paternalistic attitude the CWB had when it came to knowledge and sensitivity of what was happening back on the farm.

CWB Wheat Exports in million tonnes

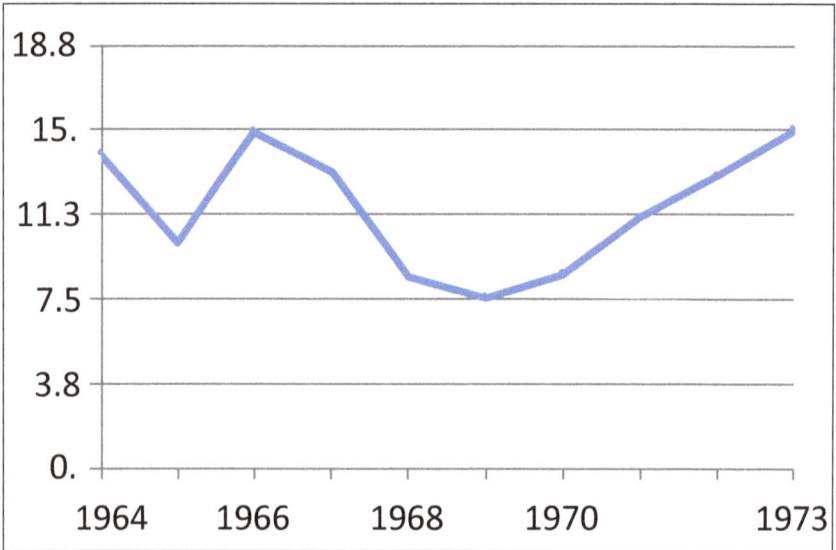

From 1968 to 1970, annual wheat exports were half of what they were in 1966 and then again in 1973. Stocks of Canadian wheat were sufficient to export more; however, the CWB chose not to enter the market, much to the financial hardship of Prairie farmers.

In the 1980s, CWB exports of oats were reduced nearly to zero. Meanwhile, prime American buyers were complaining they could not access the high-quality oats from Western Canada. It is a mystery why the CWB was not selling oats to buyers such as General Mills. While staunchly denying farmers' ability to sell oats to the US on their own, the CWB was not making any sales on their behalf.

Denying farmers the ability to market their grain to a variety of competitive purchasers suppressed the management skills of Prairie farmers. Farmers became lazy marketers as there was no opportunity for aggressive and informed farm managers to use their knowledge to market grain. It was difficult to grow a farm business when you could not budget or manage cash flow from wheat sales, nor could you plan your deliveries due to the quota system. Neither could you calculate the price you would receive until after the pool account was closed sixteen months after the wheat was harvested. Prairie farmers were treated with servility, told by the desk-ridden bureaucrats at the CWB in Winnipeg that they knew more about marketing farmers' grain than farmers did themselves. Marketing wheat is now a profit centre for informed Prairie farmers, just as it always has been with canola.

The grain-marketing-and-transportation system collapsed in the 1970s because there had been no significant investment in infrastructure by the railroads or grain companies for decades. The railroads were losing money on every bushel of grain they hauled due to the Crow Rate. The farm lobby to preserve this fixed rate was led by the far-left NFU and the three Prairie Pools and supported by the CWB. When the WGTA was passed in the 1980s, the railroads began to invest in improved grain-transportation facilities.

Notably, the CWB controlled the movement of grain through the country grain elevator system and port terminals. The grain companies that invested the capital had no control over the flow of product through their own facilities. Grain companies were hindered as merchants and most opportunities to profit from wheat sales were taken away from them. They operated as warehousemen with a handling-and-storage tariff negotiated between the Western Grain Elevator Association[57] and the Canadian Grain Commission (CGC), another federal government agency. In this environment, there was no incentive for grain companies to invest in new, larger, upgraded facilities to handle grain.

The CWB always maintained that since they were the only source of Canadian wheat on the world market, they had the power to get the best price. However, they would never open their books to prove this. Since the CWB had a monopoly, there were no competitors by which CWB's performance could be measured. International traders knew what price the CWB was selling for as the grain trade is a small global community. Only Western farmers did not know how well the CWB was doing.

---

57 A group that was heavily influenced by the Prairie Pools.

Grain companies, including international merchants, were fearful of facing retribution if they criticized the CWB. For example, if you were a Canadian grain company that pointed out flaws in CWB performance, you could find you mysteriously did not seem to get the number of railcars you were entitled to receive. Or railcars would not be directed to your export terminals. Or vessels would not be directed to your export terminals.

In later years, as the world grain market became more complex, the CWB often used third-party grain merchants to make the actual sales. These parties knew exactly what prices the CWB was getting for farmers. If they revealed that the CWB performance was not what it was claimed to be, they could find themselves off the CWB list of accredited exporters, as they were called. Now, these merchants are retired and are free to speak about what an ineffective marketer the CWB was.

In the waning years of the CWB, independent grain-market analyst John De Pape from Winnipeg closely analyzed CWB performance and released the information in a series of newsletters called *The CWB Monitor*. He used the newly released CWB Annual Reports to expose the many blunders and flawed policies of the CWB.

After the CWB was privatized in 2015, grain companies gained total control over facility operation and product flow. What followed was tremendous capital investment in grain-handling facilities, both on the Prairies and at the coast, by both domestic and foreign entities.

There is no indisputable evidence that the CWB single desk made more money selling farmers' wheat than farmers could have made for themselves in a marketplace of multiple buyers competing for their wheat. Everything was secretive. The interminable era of the great charlatan of Canadian wheat marketing was over.

The collective marketing of wheat on the Prairies, as regulated by the CWB, an agency of the federal government, was a failure.

A lie will travel 1000 miles while the truth is getting its boots on.

– E.A. Partridge Founding President, Grain Growers'
Grain Company

CORPORATE GOVERNANCE

Looking at the grain companies on the Prairies, for seventy years the industry was dominated by four farmer-owned companies with a combined market share as high as 75% of grain handled. They were all governed by farmers, who were elected by farmers. Today they no longer exist.

Elections do not often result in skilled board members, as elections are more of a popularity contest than a skills assessment. Good speaking skills, personal popularity, and a charismatic personality can often mask a lack of business acumen and corporate governance skills.

Political appointees do not often make good board members as their appointment usually reflects the repayment of a political debt, such as electoral support, rather than the specific skills of the individual. More recently, the appointment of "politically correct" candidates rather than best-qualified candidates seems to have trumped all other skills when it comes to political appointments.

Farmer-elected board members do not always make the best board members. They often do not have sufficient business or management skills to provide proper direction and oversight of large public companies. Neither are they skilled enough to deal with situations such as mergers and takeovers, which require complex and intense corporate skills. They are sentimentally too close to their organizations and often act as cheerleaders for their companies rather than making informed business decisions based upon cold, hard, objective, and strategic financial analysis.

Finally, there were the elected farmers on the board of the Canadian Wheat Board. The rump of directors supported by the hard-left NFU was there to preserve and promote the CWB single desk at all costs. There was no other objective in their minds. These directors led the CWB in an unyielding verbal battle with the federal government with no hope of discussion or compromise. They were simply waiting and hoping for a Liberal government to be re-elected to remove the threat of change at the CWB. The Harper

Conservative government in Ottawa held the future of the CWB in its hands. The group of leftist directors chose to gamble and play an all-or-nothing game with the federal government. Once the Harper government received a majority, the CWB was immediately changed. The CWB no longer exists.

This is not what corporate governance is all about. The only business plan this group had for the CWB was to hope the Liberals would get re-elected.

Hope is not a strategy.

–Vince Lombardi

Corporate governance is defined as the system of rules, practices, and processes by which a firm is directed and controlled.

Fiduciary duty is the legal responsibility to act solely in the best interest of another party.

Directors on boards must fully understand and be guided by these responsibilities and the legal implications of not doing so.

Boards of directors oversee the operations of a firm by hiring the Chief Executive Officer (CEO), approving the annual budget and the business plan, assessing risks the firm may be exposed to, and monitoring ongoing operations to ensure that affairs are carried out in the best interests of the firm.

The considerable responsibilities of a board member are not to be taken lightly. A broad set of business and management skills are highly desirable to qualify a board member to properly execute the duties of a director.

Boards of firms employ a tool entitled a skills matrix to attract candidates who have the necessary mix of skill sets to carry out the duties required. For example, it is highly desirable to have board members with accounting and legal skills. Board members should have good management and communication skills and be sufficiently knowledgeable about the business to be able to question the decisions of the CEO and senior management team. They must be confident and bold enough to challenge actions or activities they are not sure of or ask for more information on matters they don't fully understand.

Often board members may be selected because of their knowledge of the business the firm is in or the associated businesses, such as upstream or downstream partners of the said firm. The same can be said of candidates who are informed about the customers or competitors of the firm.

The selection of board members is so important to the success of a firm that a board committee will often be established to conduct a search when a directorship becomes vacant. The committee may employ an executive search agency to recruit suitable candidates, using the skills matrix to identify specific talents sought.

The Alberta Wheat Pool and Manitoba Pool Elevators, later Agricore, failed to deal with the weak capitalization position on their balance sheet. They were carrying millions of dollars of farmers' patronage dividends as an asset on their balance sheet. This was a liability since these patronage dividends eventually had to be paid out, like a debt to farmers. As payout of the dividends was coming due, Agricore had to look to debt financing to pay them out. Their balance sheet could not handle this amount of debt, and that led to their dire financial predicament when they came to UGG to talk merger. This was a major financial oversight by the boards of AWP and MPE and later Agricore. It proved fatal to the company.

The Saskatchewan Wheat Pool successfully dealt with their patronage dividend obligations to farmers in the same manner as UGG did. They took the company public and attracted outside investment. They were in a strong financial position but unfortunately an out-of-control CEO went on an unchecked spending spree on many questionable investments, and this began the financial demise of SWP. Where was the SWP board? They obviously did not have the knowledge or confidence to challenge his actions and terminate his employment earlier. This proved fatal to the company.

UGG had dealt with their patronage dividend payouts by going public in the early 1990s. Their balance sheet was strong, but the company was small. They were seeking growth opportunities or a merger to gain a stronger market position. Since the Agricore balance sheet was so weak at the time of the merger, the pro forma finance estimates for the new Agricore United (AU) were weak, even after factoring in the stronger financials UGG brought to the table. The short crop of 2002 and lack of planning for this reality by the new Agricore United board and senior management team resulted in a weak start for AU. The board and management of the new AU were so ebullient at having completed the merger that they felt their new market dominance would carry them through. Their feeling was that if there was a short crop, all grain handlers would suffer equally and since they were the largest, they would be fine.

The sudden resignation of the chairman at the merger meeting resulted in factions developing on the board, with no clear leadership emerging. The resulting divisiveness, lack of leadership, and multiple chairmen in a short period did not provide the strong board leadership necessary to guide the evolution of the new company, the merging of two corporate cultures, and the weathering of financial losses from the short crop of 2002.

The newly invigorated SWP, led by CEO Mayo Schmidt, totally surprised AU. SWP, from a weak financial position, launched a hostile takeover of AU. As AU was attempting to fend off SWP by making a deal with Richardson, SWP CEO Schmidt went behind the scenes and cut a deal with Richardson. Even though AU had a stronger balance sheet than SWP, and AU was now the largest grain handler on the Prairies, they did not fend off the much weaker SWP. AU had nobody left to save them. This proved fatal to the short-lived AU, which had deep roots going back to 1906. This was the end of the farmer-controlled grain companies on the Prairies.

Weak corporate governance by the farmers on the boards of the above entities resulted in their demise and exit from the grain-marketing-and-transportation landscape in Western Canada.

The Directors College, and Institute of Corporate Directors, offer excellent training programs for Corporate Directors.

THE INTERACTION OF BOARD AND MANAGEMENT

The board of directors hires the chief executive officer (CEO) and approves the budget and business plan for the company. The board then monitors how well management is doing. Management, led by the CEO, is responsible for day-to-day operations. The board will usually set a maximum of how much money the CEO can spend on a project or investment before having to come back to the board for approval. This is the manner the board uses to control high-cost expenditures authorized by the CEO alone.

One of the most frustrating annual issues as a board member was that of compensation for company personnel. Most difficult was remuneration for senior management and especially the CEO. Every year management would submit a report to the Compensation Committee of the board requesting higher and higher pay for senior executives. There would be long reports from consultants, paid for by shareholders, explaining why, in their opinion, senior management should be paid more—at times much more. This relates to the problematic situation in today's corporate world of astronomical CEO salaries.

During my time at the CWB, we never had these continuous salary-increase requests from management that we had at both UGG and the Credit Union Deposit Guarantee Corporation. This speaks much to the character of the individual in the position of CEO.

The CEO works closely with the board as a conduit to the personnel employed by the company. These self-serving requests by the CEO are awkward to deal with. It seems that as soon as someone gets to this position, they begin to think they are irreplaceable and worth much more than they really are. Sometimes a company is fortunate to have a CEO that really stands out in ability. This was the case with Mayo Schmidt. With Schmidt as CEO, the much-underestimated SWP successfully took over the much larger Agricore United. The same cannot be said of AU executives whose company was in a stronger financial position yet lost the company within five years of the UGG/Agricore merger.

> It takes a strong board to challenge the ego of a self-important CEO.

Management prepares a tremendous amount of material that board members must read to stay informed about the operations of the company. This is time-consuming for all parties. Directors must be wary of the way this material is presented. Crafty wording and long passages of irrelevant material can mislead directors and distract them from issues that the board should be better informed of, having been buried in lengthy, less relevant text. Seasoned, well-trained directors are aware of this and have learned to extract relevant information from management through clearly directed questioning.

THE NEW SYSTEM OF GRAIN MARKETING, HANDLING, AND TRANSPORTATION

Since the Crow Rate and CWB's marketing system have been changed, billions of dollars have been reinvested by private investment in Prairie grain infrastructure. New concepts such as unit trains and loop tracks have been introduced, making the system much more efficient. Grain companies responded by building large grain terminals in the country with unit train car spots to gather farmers' grain. For the first time in decades, there has been new private investment in grain terminals on the West Coast.

The entire system from farm gate to coastal terminal spout has been modernized and now has sufficient surge capacity to handle large crops. Reformation of the system to one that is market-driven has been an unmitigated success on all accounts.

Prairie grain production has increased, as farmers are now able to sell their grain whenever and wherever it best suits their individual situations. The profitability of Prairie farms has increased, as have asset values.

The federal government has engaged the services of Quorum Corporation to annually monitor the performance metrics of system throughput. Hopefully a reoccurrence of the collapse of the system as we experienced in the 1970s will never happen, with market forces now guiding system evolution.

THE FARM OF THE FUTURE

The changes that have occurred on the Prairie landscape over the past fifty years have been nothing short of incredible. Not only on the Prairie farm, but in the system of marketing, handling, and transporting grain and oilseeds.

The well-managed, well-capitalized, multigenerational family farm will likely continue to be the backbone of the Prairie grain farm structure. The skill set necessary to successfully manage these farms will be considerable. Crop choices must be made with consideration for profitability, soil management, conservation, and long-term sustainability. Crop varieties must be chosen from the multitude of varieties—GMO, hybrid, or otherwise. Decisions must be made as to crop inputs—fertilizer, herbicide, and fungicide applications. These agronomic decisions must be researched well to create the platform for production of a good crop, providing it rains at the right time.

The seasonal window to grow a crop on the Prairies is short. These large farms operate multiple planting units, and each unit must be served with the proper seed and fertilizer in a timely manner to keep the units running as smoothly as possible. In the fall harvest season, multiple combine harvester units are taking the crop off, and several trucks and grain carts are necessary to transport the high volume of grain harvested from the field to farm storage while maximizing combine harvester efficiency. All of this requires a manager with good logistical skills to keep everything running efficiently.

The evolving field of information technology (IT) is providing new means for farmers to monitor and analyze farm performance. The challenge will be to manage the reams of information captured by modern farm equipment. This data can be used to monitor and analyze both equipment and operator performance on a real time basis. Combine harvesters construct yield maps which illustrate the crop yields in various parts of the field. Crop planters can then use this data to apply variable rates of inputs in the upcoming year to enhance the crop performance of lower yielding areas.

Autonomous farm equipment is on the horizon. Early work is in progress on the use of drones for crop spraying. Drones are already in use to visually monitor crop growth and performance.

Modern farms generate millions of dollars in annual cash flow. Capitalization of these farms is measured in tens of millions of dollars. Complex and expensive farm equipment must be understood and properly utilized to complete all necessary field operations. The successful farm manager must be sufficiently financially literate to manage the financial element of the farm.

As each farm has several skilled employees able to operate the complex farm equipment of the day, personnel management is an additional skill required. And over the long days of winter, market intelligence and considerate application of the information available can result in good marketing, contributing to the profitability of the farm.

RISKS TO THE FAMILY FARM

The land base is one of the most critical elements of a grain farm. Expanding farms need to acquire more land to keep growing. Having land in contiguous parcels minimizes the time necessary to transport large equipment from field to field. In my opinion, a singular threat to the Prairie family farm is not so-called large, corporate farms, or investor-based land purchases. It is the Hutterian Brethren, a religious-based communal sect. They call their communes "colonies," and they have been expanding aggressively, particularly in Alberta. Once a parcel of farmland is purchased by a Hutterite colony, it is not likely to ever come on the market again, because they do not sell land. They have a large base of inexpensive colony labour, and they utilize modern technology to operate their farms. If you want your farm to survive on the Prairies, you must be able to compete with the Hutterites.

Successful farms today are large businesses and require a high degree of financial literacy. A large and stable land base is necessary. It takes more than one generation to assemble such a land component. This process involves intergenerational farm transfer and succession. As farms grow larger and more complex, with large financial commitments, it is important that the generation about to take over the farm has the necessary skill sets to do so. Also, the development of a clear plan for management transition and provisions for non-farming children will prepare the incoming generation with the knowledge of just how the plan will roll out. This is the responsibility of the senior generation, to provide for this type of harmonious transition and know when to step back and let the young generation take over.

Another evolving risk that is becoming increasingly alarming to farmers is the public's growing awareness of the environment and conservation, along with carbon emissions and the effects on climate. Increasing attention is being paid to the role agriculture plays in these issues, and environmental extremists are calling for restrictions and regulations to be placed on farms. The fact that we can feed a world of eight billion people attests to the success of our agricultural industry. Restrictions on the ability of farmers to continue producing in the current fashion will lead to lower production and higher prices.

This situation is becoming increasingly highlighted primarily by our science-illiterate urban politicians who are utilizing the political correctness of this issue to get re-elected. They fail to realize that an affluent society can only exist if there is an abundant supply of affordable and nutritious food. Once the politicians turn it over to bureaucrats to regulate, we can be sure to see more rules and regulations, licensing, permitting, and compliance inspections hindering the production of food. This will make it more difficult and costly to run a farm. The public is now totally removed from agriculture and has largely forgotten where their food comes from, and how relatively inexpensive it is.

One can see how difficult it is to farm in Europe with all their rules, regulations, and time-consuming reporting procedures administered by multiple and overlapping bureaucracies. The disturbing trend is that this type of hyper-regulation is making its way into North America.

THE ELEPHANT IN THE ROOM: OUR CLIMATE

He that will not apply new remedies must expect
new evils for time is the greatest innovator.

–Francis Bacon

Though not directly related to the theme of this narrative, the entire issue of agriculture and our climate bears mention as a future risk for farmers. Perhaps not so much the actual changes that are evolving, but the way politicians and bureaucrats are dealing with them. This brings us back to the entire issue of political and bureaucratic overreach and regulations that adversely affect our ability to produce food.

Farmers must live in harmony with the weather, climate, and environment. We've done so since the Neolithic Revolution. Survival of human beings depends upon abundant food production by the agricultural community. In large part, agricultural production around the world has evolved to its current state by adapting to the weather and climate of the region. As our population increases, more and more stress is placed upon our food production and supply chain.

With the increase in global population to over eight billion people, and the increasing need to utilize our resources efficiently to feed, clothe, and shelter this growing population, strain is being placed on our environment. Over time this could affect our climate.

> Mankind's ability to innovate is limitless, and we should be encouraged rather than penalized for doing so. Innovation comes from individuals and the scientific community, not bureaucrats. The job of government is to facilitate innovation.

It is very important to keep our understanding of this evolution based upon science, not ideology. Unfortunately, our politicians of the day have turned this into a political discussion, often in conflict with the science of the matter. Policy development is being guided by populist appeal to voters rather than reasoned debate and expert input. Big companies with scientific knowledge are treated with distrust rather than as partners in developing solutions. Society, industry, and the scientific community must be encouraged to seek new, collaborative solutions to address our environment and protect it from harms such as carbon emissions, pollution, and other damaging effects.

When the need arises, humans in general have an innate resourcefulness in their ability to solve problems, rising to meet the occasion. One must only look to World War II, and the innovations that arose because of the extraordinary needs to fight the war. In short order, after the attack on Pearl Harbor that brought Americans into the war, American factories retooled from manufacturing consumer goods to manufacturing armaments necessary for the war effort. Governments and industry worked in harmony for a common goal.

Medical innovations during World War II included flu vaccines, penicillin, and blood plasma transfusions. Mechanical and electronic innovations included the jet engine, computers, and radar. Nylon and synthetic rubber were in their primitive stages when the war began, and the demand and evolution of these products was accelerated during the war. Nylon was needed for parachutes, and synthetic tires were developed since most rubber trees were now in Japanese-occupied areas of Southeast Asia.

Arguably the greatest innovation during the war was the atomic bomb and the creation of nuclear energy, a new form of power. From a series of theoretical mathematical formulas calculated by German Jewish physicists in 1940 and through the Manhattan Project in the US, these scientists were able to develop an atomic bomb by 1945. This weapon led to the end of the slaughter of World War II when the Americans dropped these bombs on the Japanese cities of Hiroshima and Nagasaki that August.

Necessity is the mother of invention.

Over the last thirty years or so, look how far the computer has come, as well as electronic communications. Our lives have been changed dramatically forever. Now we are seeing the development of AI (artificial intelligence) as computers with vast information communicate with each other to solve tomorrow's problems.

In my opinion, this is the type of innovation that must be unleashed to solve our problem of atmospheric carbon. Advances are being made in sequestering carbon underground and in the ocean. This type of research must be encouraged and accelerated. Carbon offsets are being identified and fostered as new techniques to reverse the damage caused by excess carbon pollution. Governments must do more to promote these types of research by private enterprise, not penalize the creators of these beneficial products such as is now going on with glyphosate in the US, or genetically modified organisms (GMO) in the EU.

In my view, glyphosate is a tremendous breakthrough in crop-protection technology. It is economical and environmentally friendly.[58] From my experience on our farm, the use of harsher and less environmentally friendly crop-protection products was replaced by glyphosate. Also, the use of glyphosate for weed control replaced much soil cultivation, allowing for more carbon sequestration in the soil, and protection of the soil by not exposing it to wind and water

---

58 According to the United States Environmental Protection Agency, "The draft human health risk assessment concludes that glyphosate is not likely to be carcinogenic to humans. The Agency's assessment found no other meaningful risks to human health when the product is used according to the pesticide label." EPA Press Release, December 18, 2017.

erosion. Less cultivation means less use of diesel fuel and lower carbon emissions.

Instead of embracing the development and acceptance of these types of advances, pockets of litigious citizens in the US are suing manufacturers of glyphosate. Large pharmaceutical companies, who have the resources to pursue innovative technologies, are forced to utilize their resources to defend themselves from a hostile public determined to penalize them for being innovative and working to find new solutions to address our food-production and environmental-protection issues.

Similarly, some groups are rejecting GMO seed production, which introduces traits into seed that allow for less use of harsh pesticides. GMO also allow for the introduction of traits that are beneficial to humans. This technology could also possibly introduce traits that would lower carbon emissions.

Some countries around the world still have bans on some GMO seeds. It is the height of hypocrisy for the public to demand cheap, plentiful, healthy and environmentally friendly food, yet punish the companies that develop new techniques to enhance safe food production and meet those very demands.

We can feed the world today largely due to advancements in seed technology, pesticides, and fertilizers, primarily those of the past seventy years. In 1950, the average yield of corn was 40 bushels/ acre in the US. It is now over 170 bushels/acre. This is mainly due to better seed varieties, hybridization of seed, the application of GMO technology, and the development of glyphosate.

The holy grail of seed technology would be to develop wheat, corn, and canola varieties that can fixate their nitrogen needs from the atmosphere. This would reduce the carbon footprint of food production by having less need for petroleum-based fertilizers manufactured from fossil fuels, which cause carbon emissions. Soybeans, alfalfa, lentils, and peas all can fix nitrogen from the atmosphere. These crops are called legumes.

The holy grail of carbon emissions would be to take carbon from the air and turn it into a form of reusable energy by creating hydrocarbon fuels from nothing but sunlight and the air around us.[59] This would become a closed loop of carbon use and recycling. This technology is in its very primitive stages. Some say it will never be feasible.

That is what was said of atomic power in 1940!

The innovations we have witnessed since World War II were developed without the assistance of computers and AI. The current use of computers, and further development of AI, will assist in the development of new innovations.

> *The possibility that we may fail in the struggle ought not to deter us from the cause we believe to be just.*

> –Abraham Lincoln

---

59 Verhovek, Sam Howe. "Can Technology Help Fix the Climate Crisis?" *National Geographic*, volume 244. Issue 5, November 2023: p. 95.

EPILOGUE

A prosperous society is built upon a successful and well-developed agricultural foundation. Civilization does not move on to greater accomplishments if man must spend most of his daily efforts searching for food and shelter.

In 1943, American psychologist Abraham Maslow identified the needs of human beings and measured the advancement of society based upon what level of need a society could provide its citizens. Less advanced societies have difficulty providing sufficient food and shelter for their people, and that is still the case in many of the poorest countries in the world. Wars in some parts of the world cause people to spend most of their time dodging bullets and rebuilding after the destruction caused by warfare. These populations are still spending most of their resources on basic needs such as food and shelter, safety, and security.

On the other hand, advanced societies provide all levels of individual need until individuals reach the fulfillment of self-actualization. We have reached and sustained this highest level in North America.

Maslow's hierarchy of needs is divided into five levels of need. The first level includes physiological needs such as food, water, warmth, and rest. The second level includes safety needs such as safety and security. These are the basic needs of mankind. If a society must spend most of its time producing or searching for food, or engaging in warfare, it cannot move on to achieve higher pursuits in life, such as music, the arts, sports, and individual activities such as hobbies.

As societies develop, the third level of need includes a sense of belonging, love, intimate relationships, and friendships. Once you have achieved a level of having sufficient food, proper shelter, and a secure environment, people have time and energy to pursue partnerships, family activity, and friendships.

The next level is esteem, prestige, and individual feeling of accomplishment. This includes a sense of satisfaction with one's accomplishments.

The fifth and highest level of societal achievement is self-actualization, achieving one's full potential, including creative activities. Writing this book could be considered self-actualization. If you are hungry or cold, or living in a war zone, or in poor personal relationships, or unhappy with your status in life, it is not likely you would have the physical or mental comfort and dedication to spend time and resources writing a book.

Maslow's Hierarchy of Needs

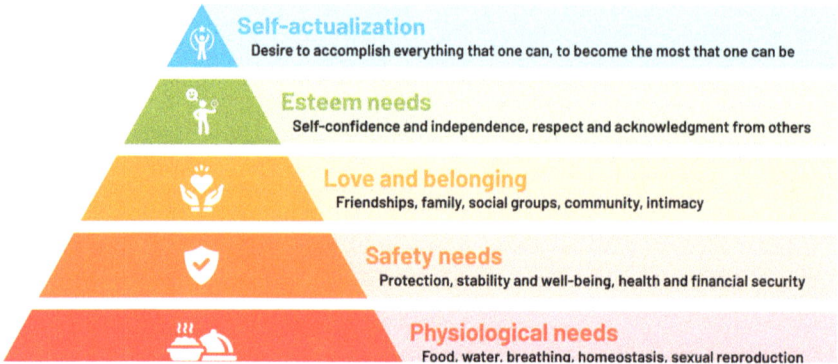

Self-actualization
Desire to accomplish everything that one can, to become the most that one can be

Esteem needs
Self-confidence and independence, respect and acknowledgment from others

Love and belonging
Friendships, family, social groups, community, intimacy

Safety needs
Protection, stability and well-being, health and financial security

Physiological needs
Food, water, breathing, homeostasis, sexual reproduction

The main point is that the less time a society must spend feeding itself the more time is left to pursue other more advanced accomplishments. North American and Western European societies have the luxury of having sufficient resources to achieve all five of Maslow's hierarchal needs. This is largely because an advanced society spends a very low percentage of income on food, leaving more resources available for other endeavours.

North Americans spend 11% of their disposable income on food while Europeans spend 14%. Contrast this with many poorer Asian countries that spend 40% of their income on food and many severely deprived African countries that spend 50% or more of their income on food.

This explains why North American and Western European societies have such a high standard of living. The arts, music, sporting activities, and individual advancement through higher education are all well developed here. We have highly advanced medical technology, dental care, and longevity unknown to past generations.

Contrast this to people in developing countries in Africa and Asia who spend 50% of their daily activity gathering enough food, and then spend additional resources protecting themselves from warring enemies trying to steal from them or harm them. This does not provide much opportunity to learn music or become a professional athlete.

Hence, an advanced society is one that has an efficient, economical, and highly developed agricultural industry that provides ample nutritious food at low cost. Our industry in North America has been so successful that now many people take their plentiful food supply for granted.

The risk we have exposed our affluent society to is the introduction and intrusion of regulatory government policy that makes food production more expensive. Needless regulations that cause us to utilize more resources for food production than necessary take us backward in societal evolution, leaving less time and fewer resources for personal advancement and achievement.

We must be wary of this encroachment of needless government policy and regulation that adds to the price of our food. The world has never experienced as efficient an agricultural industry as we now have. We have a plentiful supply of nutritious food produced at low cost. Let's keep it that way.

Modern prairie grain terminal in the sunset

EVENT TIMELINE

1867 Ontario, Quebec, Nova Scotia, and New Brunswick unite through Confederation to form the Dominion of Canada.

1871 Prime Minister Sir John A. McDonald brings British Columbia into Confederation and promises to build a railroad across the nation to link the West Coast to Upper Canada.

1885 The Canadian Pacific transcontinental railway is completed, passing through the southern Prairies.

1890–1914 The Western Canadian Prairies are settled by immigrants from Europe and the wheat economy emerges.

1906 Western farmers band together to form The Grain Growers' Grain Company, the first farmer-owned grain co-operative in Western Canada.

1917 The Grain Growers' Grain Company amalgamates with Alberta Farmers' Cooperative Elevator Company to form United Grain Growers Ltd.

1919 Under the War Measures Act the first Canadian Wheat Board is formed to control wheat supply, trade, and price during World War I.

1920 This CWB created under the authority of the War Measures Act is disbanded.

1916 Two railroads, the Canadian Northern and the Grand Trunk Pacific, default on their loans as they race each other to build a northern railroad across the Prairies and through the Yellowhead Pass on to Vancouver.

1925 The federal government formally takes over the assets of the Canadian Northern and Grand Truck Pacific and creates the government-owned Canadian National Railroad. The Crowsnest Pass fixed-freight rate is enshrined into legislation and the railroads are bound in perpetuity to haul grain at a rate fixed at 1897 prices.

1913–1931 At the urging of Western farm groups, the federal government builds several large concrete grain terminals, initially in Saskatoon, Moose Jaw, and Calgary.

1923–1924 The three Prairie Wheat Pools are formed and together they form the Central Selling Agency to market farmers' wheat through a pool.

1930s The Great Depression. The three Wheat Pools overpay for farmers' grain and their respective provinces provide them with loans to keep them operating. The R.B. Bennett federal government takes over the surplus supplies of wheat from the CSA, which is no longer financially viable.

1935 A voluntary Canadian Wheat Board is formed to sell farmers' wheat.

1943 During World War II, with instruction from the British government, the Canadian Wheat Board becomes mandatory to control price, supply, and trade of wheat. Again, this is done under the authority of the War Measures Act.

1949 The marketing of barley and oats are placed under CWB authority by Order in Council. The CWB now controls all sales and movement of Prairie wheat, barley, and oats.

1972 The Great Soviet Grain Robbery occurs, and surplus supplies of world grain are substantially diminished. The price of wheat triples in fifteen months.

1974 The federal government takes control of domestic feed grain sales away from the authority of the CWB, though the CWB later regains the power to set quotas on deliveries.

1973 A plebiscite is held, and most Prairie farmers vote to keep rapeseed (canola) marketing on the open market and out of CWB control.

1973–1979 A surge in world demand and trade in grain finds the Canadian grain marketing and transportation system to be outdated and dysfunctional. Canada loses out on lucrative grain sales due to the inability to deliver Prairie grain to export positions at ports.

1979–1980 Don Mazankowski serves as minister of transport and minister responsible for the Canadian Wheat Board in the Joe Clark minority Conservative government.

1979 Minister Don Mazankowski establishes the Grain Transportation Authority (GTA) to allocate grain cars between the private trade and the CWB. The GTA also oversees grain-transportation issues, identifying problems and seeking solutions. The CWB retains authority over railcar allocation for CWB grain.

1983 The Trudeau Liberal government passes the Western Grain Transportation Act, and the federal government pays railways the Crow Gap, the difference between the low statutory Crowsnest Pass rates and a compensatory rate that pays the railways in full for moving grain ($646 million annually).

1989 The Mulroney Conservative Government issues an Order in Council to remove oats from CWB jurisdiction and allow oats to be sold on the open market.

1993 United Grain Growers changes its ownership structure from a farmer-owned co-operative to a public company.

1995 The Chrétien Liberal government abolishes the Western Grain Transportation Act and Prairie farmers start to pay full freight rates. Railroad branch lines are now allowed to be abandoned, and pent-up consolidation of the aged elevator system commences.

1997 UGG fends off a hostile takeover attempt by Alberta Wheat Pool (AWP) and Manitoba Pool Elevators (MPE). AWP and MPE go on to merge and become Agricore.

2002 UGG merges with Agricore to become Agricore United.

2006–2007 Saskatchewan Wheat Pool (later Viterra) is successful in a hostile takeover of Agricore United.

2011 The Harper Conservative government passes the Marketing Freedom for Grain Farmers Act, which results in the removal of the CWB's monopoly marketing of Prairie wheat and barley.

2015 The CWB merges with Bunge and Saudi Agriculture and Livestock Investment Company (SALIC) and the newly formed entity is rebranded as G3.

2015 Prairie farmers gain total freedom in how they market all their grain.

2023 Canadian Pacific Railway merges with large American railway Kansas City Southern to form Canadian Pacific Kansas City (CPKC).

2023 Global grain-trading giants Bunge and Viterra propose to merge to form a large global company rivalling Cargill in size, subject to regulatory approval in the many jurisdictions in which they operate.

PERSONAL TIMELINE

1976 Graduated from University of Alberta and began work with Alberta Agriculture.

1977 Took over the family farm at Mundare, Alberta.

1979–1980 Hired as special assistant to Hon Don Mazankowski and moved to Ottawa.

1980–1984 Worked with the Grain Transportation Authority.

1984–1986 Special assistant to Hon. Don Mazankowski at the Vegreville, AB office.

1986–1991 Worked with the Farm Debt Review Board as a farm-management specialist.

1986–2002 Served as a member of the Alberta Grain Commission.

1987–1991 Became an executive member of the Western Canadian Wheat Growers Association.

1990–2002 Elected to the board of directors of United Grain Growers.

1993–1995 Appointed as a member of the board of governors, Winnipeg Commodity Exchange.

2002 Appointed to the Alberta Financial Management Commission.

2003–2005 Was a founding partner of the Whitetail Crossing Development at Mundare, Alberta.

2003–2013 Appointed director of the Alberta Credit Union Deposit Guarantee Corporation, serving as chairman the last two years

2006–2015 Appointed director of the Canadian Wheat Board.

2015–present Farm businessman able to market my own grains and oilseeds.

ACKNOWLEDGEMENTS

Going through life there are pivotal times, events and personalities that influence decisions you make, which then chart your destiny. I would like to thank the following five colleagues who provided guidance to me as I journeyed through my professional career:

Myron Dubyk: One of my high school teachers at Mundare who encouraged me to study agriculture at the University of Alberta.

Phil Jensen: My first boss at Alberta Agriculture, who taught me many valuable career skills. He taught me to think broadly and research options, look at the big picture, find solutions rather than complain about problems, and have confidence in myself. He encouraged me to rise above community norms and paradigms that stifled progress and change in our local Ukrainian community.

Don Mazankowski: Our Member of Parliament and later a federal cabinet minister. He gave me the opportunity to learn much about how government and politics work or, more appropriately, don't work.

Paul Earl: Colleague and associate who shared his tremendous knowledge of the system for marketing, handling, and transporting Prairie grain with me. As an author and academic, Paul has published many works on the subject. I worked with Paul at the Grain Transportation Authority, United Grain Growers, and the Western Canadian Wheat Growers Association. He always encouraged me to march on in terms of being a farmer rallying for change and would remind me of the quote by John Stuart Mill: "The only thing necessary for the triumph of evil is for good men to do nothing."

Bryan Perkins: Former UGG director and successful farm businessman who served as a role model for me in corporate governance and business management. Calm and quiet, everyone listened when Bryan spoke. Witnessing the unruffled effectiveness of Bryan at a board table offered an excellent example to emulate.

A special mention and thank you to **Russ Crawford**, author and grain industry veteran, who contributed valuable assistance to this writing. From critique and content suggestions for better clarity and description, to editing, review, and fact-checking, his knowledge of the grain industry, combined with his experience as an author, provided a level of assurance to my work.

Thank you to many who assisted in preparing this narrative, providing encouragement and support. A special mention to those who reviewed early versions and provided critique, fact-checking, and motivation. These include Hubert Esquirol, Glen Goertzen, Bruce Johnson, Ron Kuhn, Greg Kelly, Blair Kjenner, Barry Lynch, and Glenn Keddie.

SOURCES AND FURTHER READING

Colquette, R.D. *The First Fifty Years: A History of United Grain Growers*. The Public Press Limited. 1967.

Fairbairn, Garry. *From Prairie Roots: The Remarkable Story of Saskatchewan Wheat Pool*. Western Producer Prairie Books. 1984.

Earl, Paul D. *The Rise and Fall of United Grain Growers: Cooperatives, Market Regulation, and Free Enterprise*. University of Manitoba Press. 2019.

Rae, J.E. *C.A Crerar: A Political Life*. McGill–Queens University Press. 1997.

Morriss, William, E. *Chosen Instrument: A History of the Canadian Wheat Board: The McIvor Years*. Reidmore Books. 1987.

Morriss, William, E. *Chosen Instrument II: A History of the Canadian Wheat Board: New Horizons*. The Prolific Group. 2000.

Earl, Paul, D. *Mac Runciman: A Life in the Grain Trade*. University of Manitoba Press. 2000.

Wilson, C.F. *A Century of Canadian Grain: Government Policy to 1951*. Western Producer Prairie Books. 1978.

Sharp, Mitchell. *Which Reminds Me: A Memoir*. University of Toronto Press. 1994.

Morgan, Dan. Merchants of Grain: *The Power and Profits of the Five Giant Companies at the Center of the World's Food Supply*. The Viking Press. 1979.

Baron, Don. *Canada's Great Grain Robbery*. Don Baron Communications. 1998

Collins, Jim. *From Good to Great*. HarperCollins Publishers. 2001.

Berton, Pierre. *The Great Railway*. McLelland and Stewart Inc. 1974.

Levine, Allan. *The Exchange: One Hundred Years of Trading Grain in Winnipeg*. Peguis Publishers Ltd. 1987.

McEwan, Grant. *The Battle for the Bay: The Story of the Hudson Bay Railroad*. Western Producer Book Service. 1975.

McLoughlin, Paul. *Ottawa Finally Kills the Crow and Farming Joins the Free Market*. Alberta in the 20th Century, vol. 12. CanMedia Inc. 2003.

Collins, Robert. *An Idea Whose Time had Arrived, the Wheat Pool Bursts into Being*. Alberta in the 20th Century, vol. 5. United Western Communications Ltd. 1996.

Byfield, Link. *The Crash of 1930 Destroys the Pools as World Marketers*. Alberta in the 20th Century vol. 5. United Western Communications Ltd. 1996.

Fulton, Murry and Larson, Kathy. *The Restructuring of Saskatchewan Wheat Pool: Over Confidence and Agency, Cooperative Conversions, Failures and Restructurings, Case Studies and Lessons from U.S. and Canadian Agriculture*. Knowledge Impact in Society and the Centre for the Study of Co-operatives. 2009.

Earl, Paul D. *Lessons for Co-operatives in Transition: The Case of Western Canada's United Grain Growers and Agricore United, Co-operative Conversions, Failures and Restructurings, Case Studies and Lessons from U.S. and Canadian Agriculture.* Knowledge Impact in Society and the Centre for the Study of Co-operatives.2009.

Fulton, Murray and Larson, Kathy. *Overconfidence and Hubris: The Demise of Agricultural Co-operatives in Western Canada.* AgEcon Search. 2009.

Crawford, Russ. *Limit Up.* Agrinomics Publishing. 2022.

Crawford, Russ. *Western Barley's Legacy: The History of the Western Barley Growers Association 1977-2022. Agrinomics Publishing. 2022.*

Miner, William M. *The Rise and Fall of the Canadian Wheat Board.* CAES Fellows Paper. 2015-2.

Bennett, Mary-Jane. *As the Crow Flies: Transportation Policy in Saskatchewan and the Crow's Nest Pass Agreement.* Frontier Center for Public Policy. 2017.

The Western Producer. Grain Transportation Issues. Special Supplement. January 28, 1993.

Ross, Jane. *Grain Elevators.* The Canadian Encyclopedia. 2006.

Vervoort, Patricia. *Towers of Silence: The Rise and Fall of the Grain Elevator as a Canadian Symbol.* Social History. Vol 39, No.77, 2006.

Vervoort, Patricia. *Industrial Building in the West: The Dominion Government Elevators at Saskatoon, Moose Jaw and Calgary.* Dalhousie University. 1991.

Everitt, John and Gill, Warren. *The Early Development of Grain Elevators on Canada's Pacific* Coast. Western Geography, Canadian Association of Geographers. 2005/2006.

Churcher, Colin. *Requiem for Government Grain Hopper Cars.* Colin Churcher's Railway Pages. July 30, 2023 (web site)

Hemmes, Mark A. Twenty-Five Years of the Grain Handling and Transportation System 1995-2020: A Time of Great Change. Quorum Corporation. 2021.

De Pape, John. The CWB Monitor. (unpublished).

Paper Wheat - the play. 25th Street House Theatre. Saskatoon, Saskatchewan. Circa 1980, National Film Board of Canada (available on YouTube) .

Blandchard, J. A. *History of the Canadian Grain Commission 1912-1987*. Minister of Supply and Services Canada. 1987.

Verhover, Sam Howe. *Clearing the Air: Can Technology Help Fix the Climate Crisis?" National Geographic*. Volume 244, No. 5. November 2023.

Canadian Wheat Board Annual Reports (selected years).

United Grain Growers Annual Reports (selected years).

Saskatchewan Wheat Pool Annual Reports (selected years).

Alberta Wheat Pool Annual Reports (selected years).

Agricore United Annual Reports (selected years).

1934: Farmers' Strikes

The summer of despair was ushered in by a meteorite which blazed across Alberta before exploding somewhere near Camrose. In many ways 1934 was the worst year thus far. The weather was better than the year before, but it had gone on for so long now—clothing got shabbier, children sicklier, and there was no end in sight. The most common method of suicide seems to have been poison. Probably it was cheaper and more easily available than firearms. Men and women drank some form of strychnine and died or were maimed.

1934 was the peak year for the bush camps, which held some 25,000 men. No organization or grievance procedure was permitted, nor was there any recreation. The men felt abandoned by God and mankind.

Tim Buck (1952: 96) says there were 189 workers' strikes in Canada in 1934, of which 109 were led by the Workers Unity League, and that 84 of these were won. There were also relief camp strikes at Drumheller, Edmonton, Lethbridge and Calgary.

There were two farmers' strikes. In March, Myrnam farmers went on a non-delivery strike against the local elevator because the agent graded all grain as "tough." The grain commission investigated and advised that the agent be transferred. The company shuffled its agents around the province and this settled the matter for a time.

In November, Mundare farmers protested against being cheated. As well as suffering unnecessarily high dockage and low prices, they thought there should be cleaning equipment in the elevators. The strike was organized by the Farmers' Unity League. Peter Kleparchuk, secretary of the strike committee, and A. J. Lesiuk, chairman, announced the strike and said that farmers had appealed to the grain commission without success. They asked all farmers and workers in Canada to support the strike.

Shortly after the strike was called in Mundare, farmers at Hairy Hill, Norma and Whitford declared their support. Pickets were set up and no grain was delivered to elevators. RCMP were sent to beat on strikers and on November 7, fourteen strikers were arrested.

222   NO STREETS OF GOLD

By November 13, the strike had spread to 24 districts including Hilliard, Two Hills, Hairy Hill, Vovrik, Royal Park, New Kiew, Bushland, Plain Lake, Innisfree, Inland.

> When scabs hauled grain, the pickets stopped them. They would lie and say they were taking the grain not to be sold, but to be milled. Though they were hauling it to sell, they had to get it ground. There were incidents when scabs were protected by the Dominion Police. Nevertheless, the pickets overturned scab wagons, wheels upward. No one could restrain the farmers' rage in the struggle for their just demands. This is our history. (Mike Novakowski. *Life and Word*. November 11, 1973. Translated from Ukrainian.)

A supporter of the strike in Innisfree said non-communists were also supporting the strike because there was no doubt about dishonest grading and low prices. The cost of production was about 65 cents per bushel for a 20-bushel-per-acre crop and the price to the farmers was about 28 cents. He wrote that it wasn't necessary to be a communist to see red. (Vegreville *Observer*. December 5, 1934)

A farmer named John Fedun said he had been intimidated when he tried to take his grain to market. The picketers had felled a pole across the road in front of his truck so that he was unable to proceed. Another farmer named John Lamash asked for and received police assistance to transport his grain across the picket line. Constable Graves of the RCMP tried to drive the wagon past the picketers and the wagon overturned into the ditch.

Picketers were charged with intimidation, obstructing police, and wilful damage. They were: Peter Kleparchuk, William Zaseybida, Dmytro Ulan, Sam Ulan, Peter Bereziuk, Fred Yaniw, Joe Osinchuk. Steve Hewko was also charged with assaulting Sergeant K. E. Heacock of the RCMP, who was also the prosecutor in the case. The defence said that Constable Graves had himself overturned the wagon in his ineptness in driving too near the ditch.

Peter Kleparchuk and William Zaseybida were sent to jail for two months. The remainder were fined $25 and costs for intimidation and $20 for obstructing police. The case was appealed and Kleparchuk and Zaseybida had their sentences lengthened to seven months. The case against Dmytro Ulan was dismissed; the rest had to pay $50.

Ukrainian Catholics obeyed their priests and opposed the strike. In December, Ukrainian nationalists held a march in Myrnam, waving the Union Jack and singing ''God Save the King.'' Accompanied by some members of His Majesty's loyal Canadian lackeys, they escorted several loads of grain to the elevator.

Mayor White of Mundare organized an anti-strike committee and swore in special constables to assist the RCMP in scabbing. The strike was losing force in Two Hills and Innisfree. There had been no organized strike in Vegreville although little grain had been delivered. The merchants said the demands were justified but they deplored the methods, as they have been saying for lo, these many years about any workers and farmers demands.

The police reported that by December 17, the situation was back to normal. Perhaps they meant that elevator companies were cheating again, because the strike didn't peter out until some days after that.

Although the strike had been defeated, the companies could no longer cheat so flagrantly. A new grading system was introduced, resulting in slightly higher grain prices.

In the same year, William Aberhart, who had been preaching on radio for some years as the head of the Prophetic Bible Institute in Calgary, read one of Major Douglas' books on social credit and was smitten. Thereafter, he preached not only evangelism but social credit.

Throughout the summer of despair, William Aberhart made promises. I don't suppose anyone believed his promises, but he spoke so well—sonorously, with confidence, and with the full force of God behind him. It sounded so easy. While everyone knew it wasn't that easy, it was nice that someone could make it sound so easy. He brilliantly and incisively denounced the society and held out a vision of another one, blessed by God, where inequality would be eliminated.

Meanwhile, Vivian MacMillan and Allan D. MacMillan charged UFA Premier J. E. Brownlee with seduction. He denied it and said it was a conspiracy to ruin his reputation. The court awarded Vivian MacMillan with $10,000 damages. She appealed for more without success. At the trial she testified she had been an employee of Brownlee and had been pursued by the Premier because his wife was sick. She was seduced in the house and in his limousine. People heard about the luxurious surroundings in which their Premier lived and contrasted it with the drabness of their own lives. Premier Brownlee, who until the trial had bombasted that it was all a dastardly plot, offered little or no defence.

In the eyes of the people he was condemned not so much by adultery as by the details of the luxurious surroundings in which it took place. Premier Brownlee told farmers to have patience; he told police to beat on the poor, but he lived in a mansion and rode around in a limousine. He resigned as premier on July 1, 1934 and was replaced on July 10 by R. G. Reid.

This completed the people's disillusionment with the UFA. MLA's no longer heeded their constituents, and spent most of their time just trying to preserve the UFA in power. The party which was to have been ruled by the people was ruled by the Cabinet, which not only didn't feel bound by the party, but no longer even bothered to explain its actions.

Things seemed to be out of people's control—perhaps it was best to

APPENDIX 2

July 9, 1979
Mundare, Alberta
T0B 3H0

Hon. Don Mazankowski
Minister of Transport
Ottawa, Ontario

Dear Don

Congratulations once again on your appointment. The ministries you assumed responsibility for are likely more than a handful, but my confidence in your abilities is shared by many others. It is an honour and a privilege for the people of Vegreville constituency to be served by a representative of your stature in the Federal government.

In reference to the CWB, grain exports and transportation, and the grain industry in Western Canada, we really have no place to go but up, and any action on your part will likely lead to improvements. The inaction and non-committal nature of your predecessor allowed the system to degenerate to its present standards, a level we cannot tolerate if we have any concern for the future of our country in world grain trade and production. The decline in world status of our grain industry is of particular concern to me. I would like to express some of my thoughts and ideas on the subject as briefly as possible.

At times in the past it was difficult to determine who was actually running the CWB, Mr. Lang or the commissioners. Public statements have been made in which one group directly contradicted the other on policy issues or problems facing the Board. Often the outcome was along the lines of the commissioners' statements. (I'm referring to comments made earlier this year on the effects open market feed grains have on the operation of the system). I feel Mr. Lang allowed himself to be overruled

...../2

- 2 -

by the commissioners, and after a while it became them, not he, who were running the Board. As a result, the operations of the Board were no longer accountable to a federally elected official. In fact, the policies of certain farm organizations were reflected through sympathetic commissioners. I do not think this is proper, or fair, or in the best interests of all grain producers in Western Canada.

The CWB's performance in both pricing and marketing over the past few years is certainly indicative of policies which perhaps were not serving the betterment of the industry. We are presently sacrificing efficiency and performance to achieve equality amongst producers. But are all producers necessarily equal? Are all areas of the prairies equally productive in terms of suitability for grain production through natural forces, proximity to markets, etc? Should areas with these natural advantages, or farm units with superior managers be forced to subsidize relatively less productive land, or land more remote from markets, or land managed by mediocre or poor farm businessmen who cannot compete under true and real market conditions?

With world grain trade becoming increasingly competitive, how much can we pay for equality when it is efficiency and incentive that will keep us responsive to changes in the trade. A monopoly situation in Western Canada will isolate our grain industry from actual occurrences in the rest of the world. This leads to inefficiency because of isolation and protection from real market forces. Subsidization is then required to support that industry and maintain it in the world market. Can the Western Canadian grain industry, and primarily the economy of Canada afford this?

I am not suggesting total abolishment of the CWB; rather, I would like to see them more accountable to the federal government and all producers, including commodity groups. I would like to see them cut their propaganda program which they so subtly maintain. They should present

.../3

- 3 -

facts, not stonewall producers with philisophical opinion. The CWB was originally intended to be a marketing agency, and there has not been legislation which gives them direct authority for policy formulation. Producers should be provided with more information, and allowed to decide what is best. Once they have the facts and understand what is happening, perhaps more producers would start to question rather than blindly follow.

Yes, we have the Producer Advisory Committee, the purpose of which is to maintain farmer input into the CWB operations. However, that group is dominated by producers dedicated to the co-operative principle, and many social rather than economic factors. I agree it was producers that elected them, but with only 42% of eligible producers voting, how representative an election was it? Apathy was the dominant factor. This is where I believe the propaganda machine spouted sufficient philosophy to ensure all its followers voted. Any organization is allowed to publicly announce its views. However, when federal funds are used, as in CWB publications (specifically 'Grain Matters'), and when federal subsidies are handed out ('The Western Producer'), perhaps we should monitor what type of information Federal Government funds are supporting.

Certainly, if more producers understood that in the last 16 years Canada has decreased its share of world wheat trade from almost 25% to 19%, perhaps everyone would not congratulate the Board so fervently for its good work. If more producers knew that in the 1977/78 crop year the price for #1 Dark Northern Spring Wheat at Portland, Oregon averaged $3.84/bu. (Canadian funds), they would realize our sole marketing agency is not commanding top prices for our wheat. In the same crop year, Western Canadian farmers realized $3.27/bu. for #1 CW 13.5% protein, basis in store Vancouver. To equal the U.S. price, CWB administration, freight, and handling charges would have to equal 57¢/bu. or more. Where then is the benefit of the Crow and the premium that Canadian wheat is supposed to obtain in the world market?

.../4

- 4 -

From January to June, 1979, farm prices for feed barley and #1 DNS, 13% protein wheat at Great Falls, Montana, just south of the Alberta border have averaged $1.83/bu. and $3.93/bu. respectively (Canadian funds). American farmers can usually deliver all they wish at quoted prices, that is, there are no quota restrictions. Just north of the border, farmers are forced to live with a present quota of 210 kg. (9.6 bu.) per acre on barley at an initial price of $69.66/tonne ($1.49/bu.), and a quota of 280 kg. (10.2 bu.) per acre on wheat at an initial price of $119.69/tonne ($3.26/bu.). This would seem to indicate we should be expecting a substantial final payment on both these crops! At the same time, we cannot forget that Montana farm prices take into account freight rates at 4 - 5 times the amount we are paying in Alberta. Can we expect Canadian farmers to survive economically, relative to American farmers given the restrictions they must cope with?

The CWB was established on the premise that a central marketing agency would command a greater portion of world trade, at a higher price. One might ask what has happened to this objective of greater market power? Canada has been losing its share of the world maket, while at times receiving lower prices than other exporters.

What has the purpose of the CWB become? It appears their main interest is to achieve total control of all grain movement in Western Canada. They use this lack of total control as justification for their poor performance.

Quotas on off-board feed grains will achieve little more than to give the CWB this total control. If a producer wishes to sell his barley for $1.40/bu. (or whatever price he can get for it on the open market) and keep his cash flowing by selling all he wants, he should be allowed to do so. Economically, a producer with this attitude is efficient. Why

...../5

305

- 5 -

should he be penalized because a poor managers' costs per bushel are higher, requiring perhaps $2.00/bu. to break even. This type of farmer looks to price pooling in hope of achieving a higher price and to ensure price parity with his neighbors. If we are to remain competitive and efficient, which type of farmer will do it for us?

Furthermore, the allegations that feed grains are plugging the system and restricting the flow of export grains is not true. The Canadian Grain Commission has a 10% restriction on the amount of storage capacity that may allocated to feed grains by any company, at any time. How can 10% storage capacity restrict the system, particularly when only 5% or 6%, and even less, of our storage capacity has been used for off-board grains in previous years.

Should quotas be imposed on feed grains, we will likely see artificially low feedmill and feedlot prices on the prairies. This local market will now become the only outlet for off-quota grain sales. It seems that groups advocating quotas have not looked this far. Efficient producers attempting to maintain their cash flow will then face even lower prices.

Groups in favor of total CWB control have also criticized producer cars. They claim producer cars have been restricting the flow of export grains. Yet, less than 1% of the grain in Western Canada was shipped by producer cars this crop year. Combine quotas on feed grains with producer cars, and we have a situation where only farmers with a large number of acres can ship a producer car. Pro-CWB control groups tend to be the smaller, less management-oriented producers. In something of a contradiction, these farmers are requesting a move which will limit their market opportunities even more than for the larger, often more progressive farmers.

...../6

Another potential downfall for our grain industry through imposition of quotas on feed grains is the possibility of losing the Eastern Canadian market. This market will likely revert back to purchasing American corn. Eastern Canadian feeders are not, and have not been enthusiastic about the CWB pricing policy based on a corn competitive price. They are much more willing to deal in a free market. A thriving trade has developed between the West and the East since feed grains were on the open market. By imposing these quotas, we are endangering a market for our grain; a domestic market which does not rely on exports.

Another policy of the CWB that I question from a logistic point of view is quotas on oilseeds to crushers. These oilseeds are transported mainly by truck. Why should the CWB attempt to control their flow when it has little or nothing to do with the elevator system or rail movement of grain?

The world market can absorb all the oilseeds Western Canada can produce. As oilseeds have been attractively priced relative to cereals during the past 2 or 3 years, there appears to be little reason why oilseeds should be held as farm stocks at the end of the year. Our rail system should concentrate on moving oilseeds. Moving several cars of rapeseed will result in a much greater export value than an equal number of cars loaded with wheat. I do not believe such a move would seriously damage our share of world wheat trade, since an effective system should be able to adequately rebound once world trade in wheat picks up. Such a move would harm little more than the CWB's control over exports.

Why is it not possible to move rapeseed to export positions by truck, free of delivery quotas? Farm prices for rapeseed usually vary between 60¢/bu. to $1.00/bu. under the Vancouver cash market, yet our freight rates are only 10-15¢/bu. Why the price spread? I believe it is because of our transportation bottleneck, and the fact that rapeseed on the prairies is worth little if it cannot be moved to a port. Why not

.../7

allow producers who wish to do so, pay trucks the required rates to move their rapeseed? A freight rate of 50¢ or even $1.00/bu. is little when compared to rapeseed valued at $7.00/bu. Allowing them to do this off-quota would result in more grain reaching the coast and ultimately exports should increase. It is cheaper for farmers to pay 70¢/bu. for freight today and sell their rapeseed, rather than store it for a year and lose 70¢/bu. in interest on stocks. I understand that truck unloading facilities are presently poor at Vancouver but the terminals could be encouraged to construct such facilities. (I believe UGG is presently doing so and Pacific Terminals can presently handle trucks).

Certainly, most of our problems evolved from the limited capacity of our system to move grain. Various groups then attempted to find scapegoats for the problems, and policy alternatives have developed. Which policies will aid us in reaching our ultimate goal? Is our ultimate goal to be the movement of grain to an export market, and developing Canada's share of world grain trade; or is it to be equity among all producers at all costs, with assurance that no profits are made by any of the parties involved?

Your efforts to develop the Price Rupert port certainly deserve commendation. However, I feel it will be difficult to move more grain to market as long as the Crow rates are maintained. There must be incentive provided to achieve action. Western Canada is losing millions of dollars, not only through lost grain sales, but through production that never occured. If producers knew they would be able to sell their grain, how much more fertilizer would they use? How much less summerfallow would they leave? These are questions which will remain unanswered as long as there is a bottleneck in grain movement.

In summary, some of the issues I feel must be dealt with as quickly as possible:

1) The power and authority of the CWB should be limited and they should be made more accountable to grain producers and yourself.

...../8

- 8 -

2) Quotas should not be imposed on off-board domestic feed grains.

3) The possibility of farmers independently trucking rapeseed to Vancouver should be encouraged.

4) The Crow issue should be resolved, and I believe the proposals agreed upon in principle by 3 of the 4 Western provinces should be seriously considered.

5) We must act quickly to raise our status as a main exporter of grain in the world before our reputation is tarnished any further. This includes expansion and upgrading of all our port facilities and main rail system. Priorities should be placed on maintenance and upgrading of main lines rather than branch lines.

Your concern for the operation and problems of our grain handling system is well known. I apologize for lack of brevity, but I am also very concerned with what is happening in our grain industry. The thoughts I have expressed are based upon my experience as a grain producer and as an agricultural economist. Should the opportunity arise, I would certainly enjoy discussing these issues further with you. Thank you.

Best regards,

KEN MOTIUK

KMM:cj

APPENDIX 3

CIRCA-

KEN MOTIUK

One of the least understood costs for a prairie grain farm is the cost of moving grain to an export market. We all know about the history of freight rates in Western Canada as it is one that is etched into our farming history. Eventually inflation eroded a fixed freight rate regime to the point that the cost of moving grain was much greater than the statutory rate the railroads were able to charge. As a result of lack of funds, grain equipment was allowed to deteriorate and grain movement to ports became very slow and haphazard. Other commodities that the railroads moved rewarded them much better.

In an attempt to buffer the revenue shortfall, governments launched ad hoc programs such as hopper car purchases and branch line rehabilitation programs. But by the early 1980's, the situation became so intolerable that the federal government finally saw fit to address the situation. Thus the Western Grain Transportation Act, an exercise of highly political compromise, was implemented. In the final draft of the Act, the Government of Canada agreed to pay to the railroads their revenue shortfall when transporting grain. As costs were to rise in the future due to inflation and increased volumes, farmers and the federal government would share these cost increases. However, deficiencies evident in the legislation from the onset are now costing farmers and taxpayers a great deal of money annually.

Critics, who were farm groups in favor of the payment going to the farmers rather than to the railroads, had argued that:

1) A payment going to the railroads serves as a hindrance to value-added industry in Western Canada. It subsidizes the export of grain out of our region, penalizing industry which would process the products here. Thus, jobs and economic activity in the region are lost since the export of grain is subsidized to the detriment of the export of processed goods.

2) Simply paying the railroads their annual shortfall
would not result in a cheaper or more efficient system
since annual bills for any rail costs would be sent
to farmers and the government. Trucks could not compete with railroads on short hauls because they would
not receive a subsidy such as the railroads were getting. Railroads would not be encouraged to look for
any more innovative or efficient ways to transport
grain, as the bill for any higher costs incurred would
simply be passed along.

It is now becoming more and more apparent these critics were right.

Simply stated, under a payment directly to the railroads a
grain farmer has to do 2 things to capture part of the subsidy for
his farm operation:

1) He has to put his grain in a railcar, and

2) He has to have it shipped out of the prairie region.

Should a farmer not fulfill these 2 prerequisites, he forfeits his
share of the subsidy. If he feeds his grain to livestock, he
looses the subsidy. If he trucks his grain into the British Columbia interior, he looses the subsidy. If he trucks his grain to
a larger elevator on a railroad mainline with a lower freight cost,
he is not rewarded for doing so.

This lack of reward for any efficiency measures introduced into
the system has resulted in the rapid escalation of our freight
bill. The annual cost of moving prairie grain to export position
is now close to $1 billion, with the taxpayer of Canada paying
about 3/4 of this cost, and farmers paying about 1/4 of the cost.
Each year (other than drought years), costs escalate as innova-
tors in the system are not encouraged to seek ways to save money.
Since farmers only pay about 25% of the cost of having their grain
shipped, they do not realize the actual cost and are thus not
concerned about introducing innovations which may save the system
at large a great deal of money.

- 3 -

The average cost of railing a tonne of grain to export position is now over $31.00 of which farmers on the average pay $8.50.
However, these costs vary greatly for individual shipping points.
For example there are low volume branchlines where the freight
costs exceed $100.00/tonne. (Yes, over $3.00/bushel for wheat).
The method we now use averages these atrocious costs into the
total freight bill. We all pay for this waste of funds through
higher freight charges at other less costly shipping points.

Farmers can no longer afford the extremely inefficient measures allowed under the Western Grain Transportation Act where
most cost increases in the system are paid for by farmers. As
the total freight bill increases, most of the extra charge is
being paid directly out of our pockets and we are seemingly po-
werless to change it under the current legislation. No one is
addressing why the costs continue to go up, and no one appears
successful at adopting any efficiency measures.

Rail freight on grain from our farm at Mundare   costs
$30.00 - $35.00/acre. What we actually pay is closer to $8.00 -
$9.00/acre.   Surely if we had to pay the total out of our poc-
ket we would look for the most cost efficient method to reduce
this freight expense. But since most of this cost is hidden in a payment made directly to the railroads, farmers don't view it as
a direct expense, and are encouraged by the government to do little if anything
about it.

Groups who successfully argued that the payment should go to
the railroads (led by prairie grain handling co-ops and Quebec
based agriculture groups) insisted that the payment could not
go to producers because:

> 1) Sending cheques to all prairie grain growers
> was too cumbersome, and
>
> 2) The resulting greater amount of livestock and
> meat produced in the west would not be able to find
> a market.

- 4 -

Both of these concerns have been addressed:

      1) Canada Special Grains Program cheques were mailed
directly to producers for 2 years in succession in
a fashion originally prescribed under the pay the
producer proposal, and

      2) The Free Trade Agreement with the United States
opens up a market 10 times our size for Western
Canadian livestock and meat exports.

Thus the 2 main concerns of these groups, whose interests are
not necessarily the same as prairie grain growers, have been
successfully addressed.

One way by which railroads can send a signal back to farmers
and grain companies with respect to which delivery points are
most cost efficient to ship grain from, is incentive rates. An
incentive rate is a lower freight rate offered at a location
where it costs the railroads less due to such factors as main
line location, high volume shipment , large car spots, or other
cost saving measures. Currently both railroads offer such
rates at various locations on the prairies. For example, both
CN and CP plan to offer lower rates in 1989-1990, by up to $4.00/t,
for shipments from high volume points. (CP currently has such a
program with savings of up to $1.25/t). CN also offers lower
rates for multiple car loadings. Currently there is a $1.50/t
reduction for 18 multiple loadings, with a $2.50/t reduction for
50 car multiple loadings, and a $4.00/t reduction for 100 car
multiple loadings, planned in 1989-1990. CN appers to be somewhat
in the lead in terms of offering various incentive rate packages.

With an incentive freight rate, farmers are allowed to capture
some of the lower costs of shipping grain from delivery points
with lower freight costs. The Western Grain Transportation Act
laid out a very elaborate and lengthy legislative structure by
which railroads would have to abide if they wished to offer incentive rates so that ultimately the railroads would not have the

power to shape the delivery system of the future. This privilege
still lies in the hands of the grain companies.

The legislation is so complex that applications of incentive
rates for grain shipments have been very limited. As farmers only
pay about $8.50/tonne or 25% of the freight rate, it is difficult
to offer a freight reduction of any magnitude that can be reflected in the farmer's wallet. The legislation as it now stands
does not allow for a full reduction of freight savings at any
point. For example, freight costs to a railroad may be $10.00/tonne
less at some delivery points under some conditions. However, this
cannot be offered as an incentive for farmers to deliver to this
point since the freight actually paid by farmers is only about
$8.50/tonne of the $31.00/tonne total cost.

Incentive rate applications on the prairies to date are com-
mendable, but the real question is "Are they large enough to make
any difference?" Since farmers pay so little of the rate anyway,
potential cost savings to them are not relatively large.

Why is everything so convoluted? Well, first of all, with a
payment to the railroads, there is no direct incentive anywhere
for either farmers, elevators, or railroads to become more efficient. Cost increases simply add up into the total freight bill,
75% of which is paid by the government. (Which history has shown
has little concern for cost over-runs). Incentive rates are the
best means by which to send a signal to farmers as to which parts
of their system are efficient, and which are not. A farmer must
see, by way of his savings, why it is better for everyone involved
if high cost facilities are rationalized, and low cost facilites
utilized. Until a larger portion of freight savings can be reflected back to farmers by way of incentive rates, this signal is
not sent.

6

If a farmer hauls his grain further to an elevator point which
is offered a freight incentive, this saving MUST be returned to
that farmer. Otherwise, he has no incentive to continue to haul,
perhaps at a greater cost to himself, to this facility which in-
curs a lower freight rate. This is the only means by which sen-
sible system rationalization can occur. Price signals through the
system myst be sent directly to the farmer. When a farmer goes
through this exercise of hauling further to an elevator that offers
an incentive rail rate, he is saving the system as a whole a
    of money and thus must be rewarded for doing so.

The drought experienced throughout most of the prairies last
year has put a great deal of economic strain on all components of
our system, and continued rationalization and cost effective decision making has been accelerated. As farmers and taxpayers we
must demand that any expenses incurred in the system are fair and
legitimate. Programs such as the System Improvement Reserve Fund
under the Senior Grain Transportation Committee must be used sen-
sibly (not politically) to assist farmers in areas where they will
be adversely affected most by branchline abandonment, elevator
closures, etc. We will all benefit if such decisions are made
prudently.

The removal of these very high cost components of the system
will bring down our arverage total freight rate from $31.00/tonne.
At the same time, since farmers at large will benefit, we must
find ways to assist those who are hurt the most by these decisions, and must truck grain the farthest. There is no better
means than substantial incentive rates at elevator points which
incur lower rail costs.

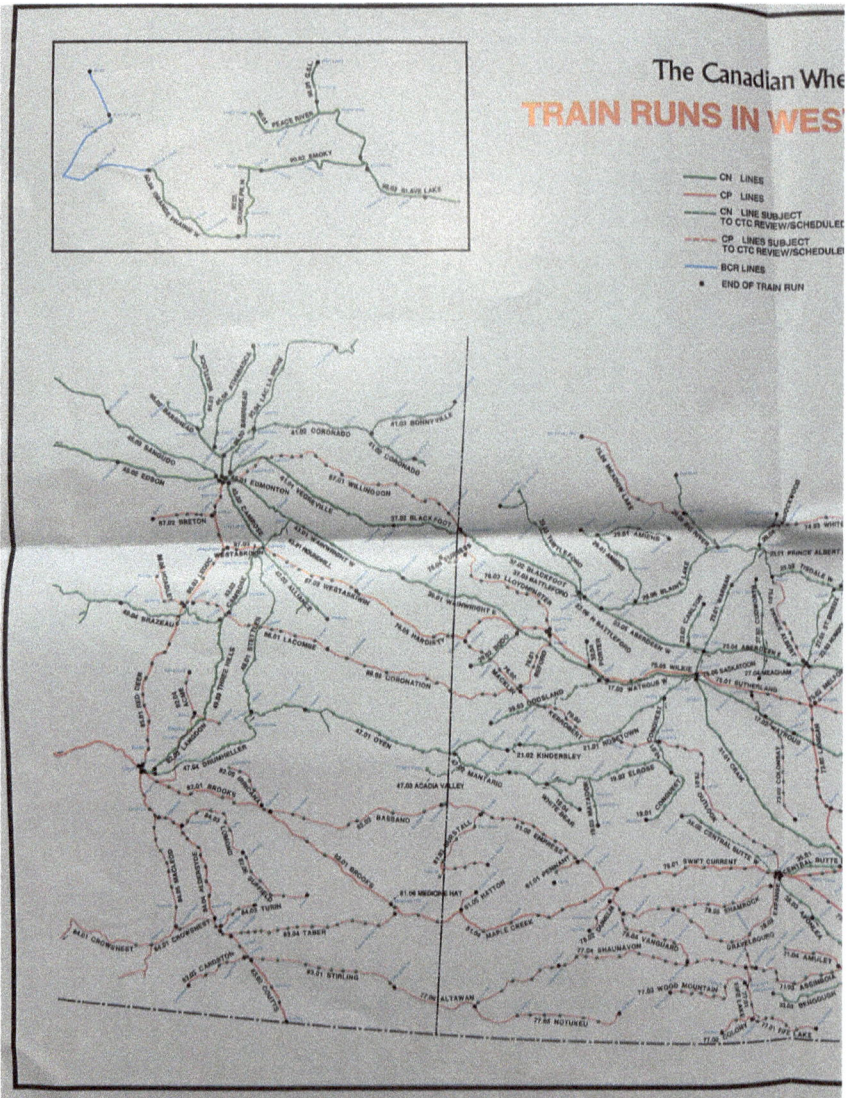

he Canadian Wheat Board

NS IN WESTERN CANADA

1935-1985

- CN LINES
- CP LINES
- CN LINE SUBJECT TO CTC REVIEW/SCHEDULED ABANDONMENT
- CP LINES SUBJECT TO CTC REVIEW/SCHEDULED ABANDONMENT
- BCR LINES
- END OF TRAIN RUN

REVISED JUNE 1985 CN GRAIN COMMUNICATIONS WINNIPEG

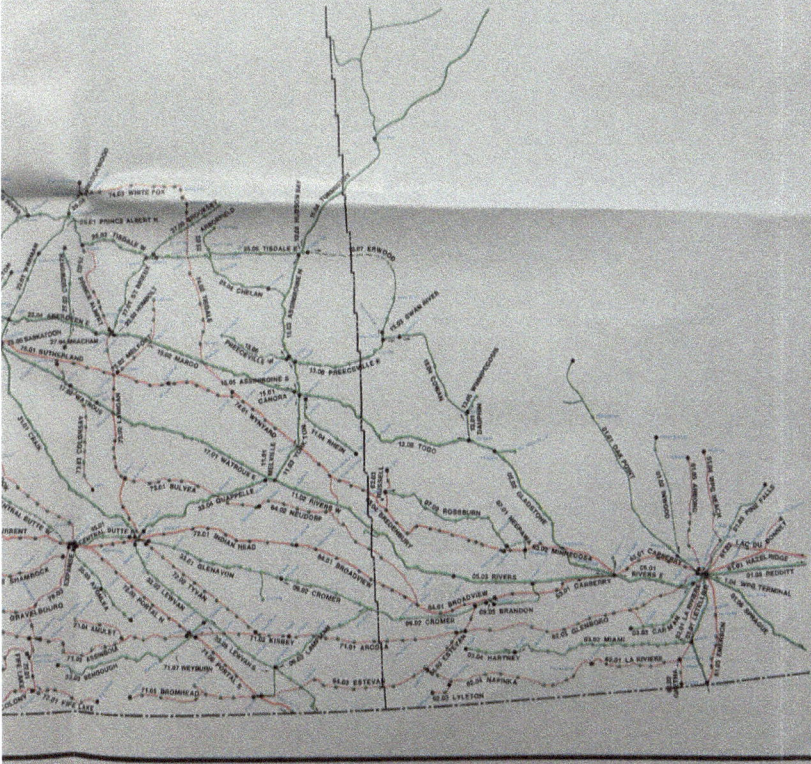

APPENDIX 5

ALBERTA-NORTHWEST OUTSTANDING YOUNG FARMER KEN AND WENDY MOTIUK WITH LOCAL COMMITTEE CHAIRMAN ED NEDZA

HERALD PHOTO BY RIC SWIHART

Mundare producer named Outstanding Farmer

By RIC SWIHART
of The Herald

There was only one nominee for the Alberta Northwest Region for the Canada's Outstanding Young Farmer competition, "but he is a dandy," one judge told an awards reception Thursday.

Ken Motiuk of Mundare, a director for United Grain Growers who farms 2,000 cultivated acres of straight grain land with his wife Wendy and four daughters, received the regional OYF trophy with congratulations from local agriculture and civic officials, provincial Jaycees representative Dan Wourms of Lloydminster, and Tim Duffin of Calgary, national chairman for OYF.

Motiuk started farming with his father in 1977, and the pair still have a working relationship.

Motiuk continues to use his double disk press drills, partly due to the uncertain economic conditions in the grain sector.

Farming in a good rainfall district, Motiuk plants all his land except one quarter section each year. Wheat is his main crop, but he "forces as much canola as possible" into the rotation. He also plants some oats.

He switched to straight combining part of the crop in 1985, and last year, did the entire farm that way. A grain drier is an integral part of harvest. It allows him to keep the combine going most of the time and still store the grain in good condition.

Straight cutting reduces his labor requirements, he said.

He markets most of his grain through producer cars, and "saves quite a bit of money."

He sells canola to a crusher, oats to the United States as pony oats, and barley to a feedlot 25 miles distant and to the British Columbia feed market.

Selling his oats and barley that way frees up delivery quota for his wheat to the Canadian Wheat Board.

His goal is to sell all the production from each year's crop. "We want to turn it into money as quick as we can."

Saving labor and soil has become a priority. To achieve that each fall, he uses a three-in-one operation. On his second cultivation, he also spreads granular Avadex and nitrogen fertilizer.

Farm management has become more critical, said Motiuk. "We find we can enhance our income at each phase of the operation if we pay attention," he said.

Motiuk is also a contractor with the Farm Debt Review Board, and serves as a commissioner on the Alberta Grain Commission.

He was nominated by the Agricultural Service Board of the County of Lamont.

The competition is sponsored by the Lethbridge and District Exhibition Association, Ag-Expo Committee, Imperial Oil, Time Air, Calgary East Jaycees and John Deere.

Motiuks named outstanding farmers

Alberta Northwest Region Jaycees presented this year's outstanding young farmer award to Ken and Wendy Motiuk from Mundare.

They are the third generation to operate the family farm.

Ken graduated in 1976 with a bachelor of science in agriculture and Wendy graduated in 1978 with a bachelor of science in nursing.

The were married in 1976, and in 1978 moved out to the farm. A mobile home was set up directly across the road from the farmstead, and they started their own site on 80 acres given to them by Ken's parents. At the time the farm had about 500-600 acres sown to crops, along with about 35 cows.

Shortly after starting farming they switched over to continuous cropping, along with more intensive use of fertilizer and herbicides. Attention was paid to crop rotations and new crop varieties were experimented with. They attempted to maximize production on a limited land base. Land purchases were expensive, and leased land was hard to come by in the late 1970's.

The farm grew a little every year, with few large moves. The conservative manner of growth eventually served them well. They financed their own seeding operations, and kept their debt manageable. Today they farm 2,460 acres of which 1,440 is rented. A home computer is used to keep financial records and do crop and cash flow forecasts.

The Motiuks and their children, Erin (12), Carlee (8), Jillian (5) and Janet (3), participate in community activities. Ken is a director of UGG, assists with the Farm Debt Review board and is a commissioner of the Alberta Grain Commission.

The program in Alberta is administered by the Lethbridge Jaycees and is sponsored by the Lethbridge and District Exhibition, Ag-Expo Commitee, Time Air Inc., Esso/Engro and Cargary East Jaycees.

The Motiuks now proceed on to the national competition in Regina, Nov. 28 to Dec. 4. The national program is a program of the Canada Jaycees and is sponsored by John Deere Limited.

APPENDIX 6

To: Ted (Where Did It Go?) Allen
   Brian (Friendly Lie) Hayward
   Terry (THE CRUSHER) Youzwa

From: Ken (Wanna Be A Golfer) Motiuk

Re: Golf scores

After a great deal of sober thought and profound soul searching, I believe it would be in the best interests of all golfers from our round on Thursday to pool our scores.

That would result in an initial score of about 96, which is really more than I could ever hope for, but feel I have a share in, since we were all in the game together. This would provide equity for everyone, as premiums from low handicaps would be blended with high scores from higher handicaps, who must be accomodated.

Advantages due to timing of execution (ability to hit the sweet spot at the exact time and location), and geographic location (distance of the ball to the hole after the first hit), would be averaged amongst all who quite evidently tried just as hard.

The result would be a more harmonius group, none of whom could support themselves on The Tour, but all of whom could wake up each morning knowing that even though they may not excel, neither would their peers. The game would continue to exist because we all love it.

Birdies and eagles (cherry-picking), are not individually allowed. They are simply a function of locational advantage (being closer to the pin), and are not putting opportunities for all others who simply find themselves further away from the premium location.

Looking forward to our final score which will be tallied after the current season is complete.

Cheers;

Note: Non-compliance with previously arranged tee times will result in a 6 stroke, individual penalty..

PS: A variation of the above would be Saskatchewan Scramble. Upon each round of hits, everyone would play the worst ball (lowest common denominator), while still trying to achieve the best score.

APPENDIX 7

THE DECLINE OF THE PRAIRIE GRAIN CO-OPS

The four farmer owned grain handling co-ops -- United Grain Growers (UGG), Alberta Wheat Pool (AWP), Saskatchewan Wheat Pool (SWP), and Manitoba Pool Elevators (MPE) helped develop Western Canada.

At their zenith in mid-century, these 4 co-ops totally dominated the agricultural industry on the prairies. They bought grain, sold coal, twine and other farm supplies and briefly sold agricultural implements. They provided livestock feed and livestock marketing facilities, as well as crop input supplies. They dominated the publishing of prairie agricultural newspapers, and were very influential players in both rural life and the prairie economy as a whole.

The 4 co-ops continued to be active in agri-politics through most of the latter part of the 20th century. They dominated farm policy and few governments challenged them. Their presence was woven into 20th century prairie life.

Now, after 100 years, their business, influence, and financial strength have shrunk. The 4 grain co-ops have become 2. Their grain handle in the past decade has fallen from over 70% to just over 50%. Both survivors trade as public companies, but have corporate governance structures that maintain farmer control at the board level.

The number of grain elevators peaked in the 1950's but were based on a horse and wagon box era built every 6 or 7 miles along raillines. Each town would have several grain elevators, usually able to load only 2 or 3 boxcars each. The economics of the system were dictated by the *Crowsnest Pass Freight Agreement* which was enshrined in legislation.

The 4 co-ops flourished in this environment. Woe be to anyone who dared to challenge the influence of the prairie grain dominators, led by the Canadian Wheat Board (CWB), the Canadian Grain Commission (CGC) and the unwavering support and influence of the 4 co-ops.

Cracks began to appear in this tidy arrangement in the 1970's as the Crow Rate caused financial stress on the railroads. Rail companies didn't want to reinvest in grain infrastructure, and were being challenged by other system participants as to whether they were complying with their statutory obligation to move grain. The system designed for horse and buggy days couldn't work in the 1970's and 1980's so grain exports and prairie farm income suffered.

In 1986 the Federal Government passed the *Western Grain Transportation Act* (WGTA) which paid the railways for their losses transporting grain under the Crow Rate.

The WGTA set the stage for a complete modernization of the prairie grain handling and transportation system. The railroads increased their capital spending on grain infrastructure but the grain companies still had the old wooden elevators.

By the early 1990's, the need for capital reinvestment was pressing. The railroads were gearing up to load 100 car unit trains. Large new concrete facilities, at a very high cost, were necessary to create an efficient logistics system.

The prairie grain co-ops had a problem; how to find money to build new, large elevators. UGG was tight on working capital and faced the pay-out of millions of dollars of debentures through the 1990's. Of the four co-ops, they had the oldest and smallest grain elevators in the system. The 3 provincial pools also had a problem. Much of their equity was tied up in retained patronage allocations to members. As the farm population was aging, they were retiring and wishing to redeem this patronage allocation.

So began the journey to reshape the grain gathering system on the prairies. In 1993 UGG, in the quest to raise capital, launched the first of a sequence of events that would cascade and play out over the next decade. UGG went public, to trade on the Toronto Stock Exchange (TSE – now TSX), with an initial public offering (IPO) at $8/share. The *UGG Act* was amended to allow for three non-farmers to sit on the UGG Board of Directors. An injection of public capital allowed for the payout of debentures coming due and converted farmers' equity into public shares that could be traded on the TSE. The injection of public equity allowed the company to start rebuilding the country grain collection system. The gauntlet was thrown down to the 3 Pools.

Next came Saskatchewan Wheat Pool. The co-op went public in 1996, with an IPO at $14/share. SWP used its public equity to reconstruct its country facilities, and invest in high risk ventures in Poland and Mexico, pork production businesses, a fish farm, farm supply outlets and feedmills among other things. The company lost of millions of dollars.

From 1996 to 1999, SWP's entry for goodwill on their balance sheet went from $3 million to $95 million, a reflection of the premium over book value paid for its acquisitions. In 2003, when SWP went through its major financial debt restructuring, **$88 MILLION IN NEGATIVE RETAINED EARNINGS** were written off their new balance sheet.

Retained earnings represent the accumulated profits and losses of the corporation since incorporation that remain undistributed to shareholders.

This left Alberta Wheat Pool and Manitoba Pool Elevators pondering their future. In 1997, they joined forces to launch a hostile takeover of UGG which was rejected. In 1997, Archer Daniels Midland (ADM) bought just under 50% of UGG for $16/share. AWP and MPE responded by merging to form Agricore Co-operative in 1998 and became the largest grain handling company on the prairies. It then borrowed money to finance new high throughput elevators.

The undisciplined spending spree and questionable financial due diligence of SWP caused it to lose **OVER $290 MILLION** from 1999 to 2003. Share prices dropped from a high of $24 to a low of $.30 in 2003, trashing farmers' retirement money. SWP was forced into a major debt restructuring in January of 2003.

To make matters even worse, part of the restructuring included a convertible debenture, which allowed for the conversion of debt into shares. With just over 60 million shares outstanding in January of 2003, by October 2004 there were over 240 million shares of SWP. If the entire outstanding debenture was converted, the market would be deluged with over 600 million SWP shares; **A 10 –1 DILUTION.**

Meanwhile, Agricore was having its own financial problems. In April of 2001, Agricore found itself in default of a loan covenant with its creditors. UGG, though not immensely profitable, now had a balance sheet that was the strongest of all the former co-ops. By merging the two companies, the plan was to be able to use UGG's balance sheet strength to offset Agricore's weakness, while paying out Agricore patronage holders through a public offering just as UGG had done in 1993. To make this successful, the new Agricore United (AU) would have to capitalize on its size and market strength, cut costs and streamline operations.

To assist in financing the merger, AU also issued a convertible debenture, the mechanics of which are similar to the one described above for SWP. If Agricore's debenture was converted to shares @ $7.50, this would dilute the current outstanding 47 million shares of AU to over 61 million shares without making the company any larger or any more profitable (other than not having to pay 9% interest on the convertible debenture).

WHAT HAPPENED?

Lets look at some other grain companies. Since Cargill and Pioneer (J.R. Richardson) are family owned private companies, no financial information is available for scrutiny. What we do know is that each of these companies are over a century old, tremendously successfull, and show no outward signs of financial stress. Also look at Weyburn Inland Terminal (WIT), a farmer owned grain company at Weyburn, SK.

Operating in the same business environment as AU and SWP, though solely situated in the southern dryland rather than diversified across the prairies, WIT has made a profit **EVERY YEAR** over the last 10 years. They have maintained a stable volume of grain throughput, and they have **GROWN THEIR RETAINED EARNINGS AT A COMPOUND RATE OF OVER 15%/AN.**

As for co-ops, lets have a look at United Farmers of Alberta (UFA), primarily a farm supply outlet in Alberta that competes with AU. UFA has been growing successfully and has doubled its annual gross revenue since 1994. It has been profitable and has **GROWN RETAINED EARNINGS AT A COMPOUND RATE OF OVER 10% per year.**

Federated Co-ops continue to grow, and has a successful retail grocery business in the Calgary market, one of the last places you would expect a populist co-op structure to flourish. FCL recently reported its 13th consecutive year of record profits. So some grain companies and co-ops have done well in the same market and same business as the former grain co-ops.

The saga continues as SWP announced its first quarter results for the 2005 year in December of 2004. They lost money again! It says it won't likely be able to pay off the convertible debenture due in 2008, and so a massive conversion and dilution of shares hangs over the business. Most dramatic was the concurrent announcement that SWP is pro-actively overturning its quasi-co-op governance structure to a standard CBCA (Canadian Business Corporations Act) format. The farmer majority on the SWP Board of Directors would no longer be. The new board would have 4 farmers, and 8 shareholder representatives from the business community. The farm community bemoans its loss of control, but creditors and shareholders were in control as soon as SWP was in financial difficulty, no matter the composition of the board.

In the mean time, how far behind is AU? It has the convertible debenture hanging over its head. **IT HAS LOST MONEY FOR EACH OF THE 3 YEARS SINCE THEY MERGED,** for a total loss of $43 million, even after receiving $14.4 million after divesting themselves of their communications division. It has lost over **$56 million** of retained earnings from its balance sheet since 2001, the last year UGG published a financial statement before the merger. This is an almost unexplainable variance from the financial success that was projected during the merger.

Every year we hear the weather is the problem. But the weather is the same for everyone, from WIT and UFA to Cargill and Pioneer. Jack Welch, the former CEO and Chairman of General Electric said he never wanted to be in a business where he could not control the main elements that influenced performance. That is hardly the case in the prairie grain handling and crop input businesses. Global commodity prices, international farm subsidies, and the weather are 3 major influences on agricultural companies doing business on the prairies. Each one of these is totally out of the control of management. That's simply the nature of the business we are in!

You cannot always blame something else for your business failures year after year. Responsibility lies at the most senior level of the company, the board of directors. Where was the board when SWP was making their dubious financial investments in Poland, Mexico and elsewhere that eventually led to their difficulties? Where was the board in 2003 when AU's 6 most senior personnel were paid almost $2 million in base salary **AND A PERFORMANCE BONUS OF $392 THOUSAND FOR LOOSING $5.5 MILLION** of shareholders' money? The Board of Directors of AU then went on to reward their 6 most senior employees with a 27% increase in base pay for 2004. AU went on to loose $13.7 million in 2004.

It's a well known fact that the industry overbuilt in the 1990's, and this contributes to its financial problems. To be profitable, the large new grain elevators must turn their capacity 10 times per year in throughput. Currently, the industry averages about 6 turns a year with an average crop. So why is AU now spending $12 million on 'grain storage expansion projects' at a time when they are loosing money and the system is already overbuild?

When UGG and SWP went public in the 1990's, critics said this was the first step in farmers loosing control of the co-op system. It turned out they were right. Only financial success would allow farmers to maintain control.

Farmers lost control of the companies when the companies ran into financial problems. Creditors and shareholders are now taking over by default.

SWP will soon be a full CBCA company. ADM does not have to make a play to take over AU - it could backslide right into ADM's arms. It did not have to be this way!

Ken Motiuk farms at Mundare, AB and is a former Vice President with Agricore United.

ABOUT THE AUTHOR

If you ask Ken Motiuk what he does for a living, he will tell you (proudly), "I'm a farmer." But when you read this book, you learn he's so much more. Ken has been a farm leader from a very young age. He has continued to improve his skills in every venture he has undertaken, and, as a result, has played a major role in the deregulation and positive evolution of the Canadian grain-marketing system. From commodity markets to transportation to government policy and farmers' rights, Ken has been at the forefront of Canadian agriculture.

Ken Motiuk is an innovator and a calculated risk-taker, preferring to be responsible for all aspects of his farm operations. Thus, his desire to escape the marketing clutches of the Canadian Wheat Board both from inside and outside the walls of 423 Main Street in Winnipeg.

Thankfully, Ken has chosen to document his experiences and memories as reminders for those who lived it with him and as a guidebook for those whose futures still lie ahead. Ken has carved his own path amid grain companies, railways, government agencies, and politicians. *Culture of Control* is his story of the journey.

Russ Crawford

www.ingramcontent.com/pod-product-compliance
Lightning Source LLC
Jackson TN
JSHW060255160425
82698JS00001B/1